David Fairhall was the *Guardian*'s Defence Correspondent throughout much of the 'Cold War' and has written extensively on maritime subjects. His previous books include *Russia Looks to the Sea* (1971), *Black Tide Rising: the Wreck of the* Amoco Cadiz (with Philip Jordan, 1980) and *Common Ground: the Story of Greenham* (I.B.Tauris, 2006).

'A masterful assessment of the fast-changing Arctic'

– John Vidal, Environment Editor, the *Guardian*

'*Cold Front* gets the story right! The story is not just about arctic sea ice retreat, but also about globalization and natural resource development driving future arctic marine transport'

– Lawson Brigham, Professor of Geography and Arctic Policy, University of Alaska Fairbanks and Chair of the Arctic Council's Arctic Marine Shipping Assessment (2005–09)

'*Cold Front* provides a fascinating and often intriguing account of the Arctic, both its history and its future. For anyone who wants to know more about the Arctic this well written and informative book provides both insight and answers'

– The Rt Hon Lord David Owen CH

COLD FRONT
CONFLICT AHEAD IN ARCTIC WATERS

DAVID FAIRHALL

COUNTERPOINT
BERKELEY

*"For my grandchildren,
and their grandma Pam"*

Copyright © 2010 by David Fairhall. All rights reserved under International and Pan-American Copyright Conventions.

First published by I.B. Tauris & Co Ltd in the United Kingdom.

Library of Congress Cataloging-in-Publication Data

Fairhill, David.
Cold front : conflict ahead in arctic waters / David Fairhill.
p. cm.
Includes index.
ISBN 978-1-58243-760-6
1. Arctic regions—Environmental conditions. 2. Ice caps—Environmental aspects. 3. Global warming—Arctic regions. 4. Climatic changes—Arctic regions. 5. Environmental degradation—Arctic regions. I. Title.
GE160.A68F34 2011
333.91'64091632—dc23

2011027260

978-1-58243-760-6

Jacket design by Michel Vrana
Printed in the United States of America

COUNTERPOINT
1919 Fifth Street
Berkeley, CA 94710

www.counterpointpress.com

Distributed by Publishers Group West

10 9 8 7 6 5 4 3 2 1

Contents

List of Illustrations	x
List of Maps	xi
Acknowledgements	xii
Foreword	xv
Preface	xvii
Introduction	xx
The Arctic Arena	1
Thickness matters	3
Coming in from the cold	5
An alarming opportunity	7
The barking dog	10
Tipping point	12
Frozen Assets	15
Russian company	16
Lamp oil and corsets	17
Cod wars	18
Striking it rich	20
A state within a state	23
Turning off the gas	24
A different ball game	25
The Law of the Sea	27
An altruistic tradition	28
Going to the ball	29

Within cannon shot	31
Rights of passage	32
On the shelf	32
Who owns the North Pole?	33
A Russian fist	37
Significant geology	39
Fishing for facts	41
Ice bears that go bump in the night	42
Rights of access	44
Cold Warfare	**47**
'The most valuable piece of real estate in NATO'	48
A helping hand	49
Island of the donkey's ears	50
The sailor Czar	51
Krushchev's herrings	53
A hint of violence	55
Wakening bear	56
The silent service	58
Polar snapshots	60
Coming up for air	62
East by North	**63**
'As plausible as the English Channel'	63
Gin on the rocks	64
Wishful thinking	67
Spared the gallows	68
Logistical nightmare	70
Filling in the blanks	72
'The man who ate his boots'	74
Dip circle	75
Creating a Victorian legend	76
Home from home	78

Ice trap	80
No English gentleman…	81
'My mission on earth'	82
Two skeletons, some chocolate and a little tea	83
Going native	84

Short Cuts — 87
First night	89
Canalside view	90
Second time unlucky	91

Bolsheviks in Cold Waters — 93
With Stalin's compliments	96
Wonderland	97
Wishful thinking	98
'Strength through joy' for Stakhanovites	99
The Bakayev plan	100
Arctic gateway	101
Atomic relations	102
Czar Bomba	103
Half-life	104
'You get used to it'	105

Breaking the Ice — 107
A nuclear pioneer	109
Out of the ordinary	110
A difficult birth	112
Atomic takeover	113
A polar giant	115
Back to front	116
Nuclear tourism	118
'Water where it didn't use to be'	119

Ottawa's bribe	121
Polar attitudes	122

North-West Passage — 125

Inuit know-how	126
An icebreaking leviathan	128
Arctic surgery	129
Pyrrhic victory	130
'Inflamed nationalists'	131
The 'arctic exception'	132
'Use it or lose it'	134
A political voice	135
Mackenzie's oil	137
What the whales will hear	138
Arctic bridge	139
Not just polar bears	140

North-East Passage — 141

Baffling statistics	143
Ice cellar	145
Late developer	146
No need to queue	147
Second refusal	149
Practical doubts	151
Turn of the tide	152
Living on borrowed time	153
Eco–tourism	155
Conquest and assimilation	157
Great expectations	158
'A resource base for the twenty-first century'	159
Ice shuttle	161
Mother of all icebreakers	162

Across the Top of the World	**165**
Changing course	169
Reality check	171
Hidden beauty	172
Meltdown	**175**
Alarm call	178
Abruptness	182
From meltdown to shutdown?	183
Possible Outcomes	**187**
Double negative	188
Dangerous waters	190
Shipping forecast	194
Northern Poll	**197**
A Chronology	**205**
Index	**211**

List of Illustrations

1. An endangered species. 4
2. The Arctic holds perhaps a quarter of the world's remaining oil and gas (Courtesy of BP). 21
3. The old coal-loading berth, Longyearbyen (Photo by author). 43
4. HMS Tireless at the North Pole-a routine operation! (Courtesy of Royal Navy). 60
5. A Victorian legend: John Franklin (Courtesy of Peter Lewis). 77
6. Nuclear icebreaker *Rossia* followed by freighter *Beluga Foresight* on Northern Sea Route, September 2009 (Courtesy of Beluga Shipping). 112
7. The double-acting ship *Norilsky Nickel* – is this the future of arctic operations? (Courtesy of Alexey Shtrek). 116
8. View from the bridge of *Beluga Fraternity*, navigating the North-East Passage, September 2009 (Courtesy of Beluga Shipping). 171
9. Will this container ship leaving Felixstowe one day head N through the Arctic, not S through Suez? (Courtesy of Port of Felixstowe). 192

List of Maps

1. A birds-eye view of the Arctic Ocean.   xix
2. Trans-Polar Drifts.   8
3. The Arctic's Natural Resources.   20
4. Legal control of the Arctic Ocean's resources.   34
5. Two Arctic Routes from the Atlantic to the Pacific.   66
6. Franklin's Last Voyage.   78
7. Siberian Waters.   94
8. Possible routes through the N-W Passage.   126
9. Routeing options through the N-E Passage.   142
10. The Arctic Option.   166
11. 2007 – A Record Year for the Summer Ice Melt.   177
12. Summer Ice Extent – 2009.   180

Acknowledgements

Many busy people have given of their time to help with this project, but my thanks go first to my wife Pamela, without whose patient support I could not have made space for it.

Those who contributed directly by (bravely) offering a specific prediction for 2040 are listed in Northern Poll. I am grateful to all of them, but several went beyond that to offer advice from their wide experience and responded generously to my innumerable technical queries – for example, Lawson Brigham and Kimmo Juurmaa regarding icebreakers, Martin Pratt on legal matters and Mark Serreze with his crucial assessments of the melting ice cap.

Among them, I am especially grateful to Peter Wadhams, who not only found time within his crowded schedule to explain the wider implications of climatic change in the Arctic, and point out valuable sources, but also to contribute his foreword.

Without explanatory maps, much of this book's narrative would have been difficult to follow. I was therefore particularly pleased to be able to enlist the services of the illustrator Kate Robinson and grateful for the excellent, lucid work she produced. Outlines were obtained from www.d-maps.com.

Within the maritime world, I turned for advice to Simon Bennett, Peter Godfrey and an old friend – David Taylor. It was David who introduced me to Peter Wright, who gave unstinting support from start to finish of this project – suggesting avenues I might explore and in particular introducing me to the *INSROP/ARCOP* studies to which he had contributed. His friendly help was immensely welcome.

Acknowledgements

Professor Lawson Brigham – icebreaker captain turned academic – alerted me to the more recent *AMSA*, a wider-ranging study which he chaired. But more than that, he gave me the benefit of his vast personal experience and steered my own investigations round several possible pitfalls.

Given the Russians' central role in all this, I could not have gone far without the access to their official thinking provided by Vladimir Vasilyev at the Central Marine Research and Design Institute (CNIIMF) in St. Petersburg. He directed me to relevant research and replied promptly to endless e-mails, dealing patiently with everything from nuclear icebreaking strategy to the eccentricities of Russian statistics.

The staff of the National Maritime Museum in Greenwich helped with other queries and the search for suitable photographic images – as did Beluga Shipping in Bremen, BP, the Port of Felixstowe and the Royal Navy. The photographer Peter Lewis, and Alexey Shtrek, from CNIIMF, kindly gave permission to reproduce their own images.

Reassurance as to the relevance of chaos theory came from the mathematician Grant Hillier.

In attempting this quick survey of the arctic future, I am of course indebted to many others – historians, scientists, economists, engineers, lawyers and journalists (especially those on my own former newspaper, the *Guardian*) – who have written with more specialised expertise on various aspects of this subject. For readers wishing to explore it further, the multinational *INSROP/ARCOP* studies referred to in the text (dealing with the North-East Passage), and the Arctic Council's wider, four-year *AMSA* assessment, provide much essential background. I also found *The Soviet Maritime Arctic* (Naval Institute Press, 1991) and *Politics of the Northwest Passage* (McGill-Queen's University Press, 1987) particularly helpful, and I referred back to the pioneering work of the late Dr. Terence Armstrong – for example, his study of *The Russians in the Arctic*

(Methuen, 1958) – and Tim Greve's official guide to Svalbard, a book I took with me when visiting the archipelago (Grondahl, 1975).

The fascinating and still controversial story of Franklin's death while trying to navigate the North-West Passage has produced a flurry of books in recent years – for example, James P. Delgado's *Across the Top of the World* (Douglas & McIntyre and British Museum Press, 1999), Ken McGoogan's *Fatal Passage* (HarperCollins Bantam, 2002), and Andrew Lambert's *Franklin: Tragic Hero of Polar Exploration* (Faber & Faber, 2009), each offering a different, thoroughly researched perspective. However, my own favourite reference for anything concerned with arctic exploration – if you can find a copy – is Jeannette Mirsky's *To the North* (The Viking Press, 1934).

Foreword

This book is a wide-ranging survey of the role of the Arctic Ocean in our present society, and the changes that global warming is going to bring to the Ocean and to our own lives. The impact of arctic warming extends far beyond the Arctic.

I first encountered this ocean in the summer of 1970, when my ship, the Canadian research icebreaker '*Hudson*', sailed through Bering Strait and entered the Beaufort Sea to accomplish an oceanographic survey of the region prior to transiting the North-West Passage and completing the first circumnavigation of the Americas. The survey was very difficult. All along the coastline, from Alaska to the Mackenzie delta, the Beaufort 'Sea' was really just a narrow slot of temporarily open water, 80 miles wide, in which we could carry out our ocean survey and tow our instruments. Beyond it, the arctic pack ice stretched across to Siberia. Yet in the summer of 2007, a similar ship would be able to operate freely across almost the entire ocean; the Beaufort Sea was ice-free and so, for the first time ever, were more than 1 million square kilometres of ocean terrain in the central Arctic.

What has caused this change? The scientific finger points at global warming. The Arctic is warming twice as fast as lower latitudes, and the higher average air temperature means that the ice grows less during the winter and experiences a longer melt season in summer, when the protective blanket of snow melts and exposes the ice surface to rapid solar-driven decay. Increased heat also impinges on the ice bottom from the ocean, because the underlying layer of warmer water that flows in from the Atlantic has heated by about a degree. Warmer water is also entering through Bering Strait, and a changed pattern of winds is moving ice out of the Arctic Ocean more rapidly, so it has less opportunity to turn into multiyear ice, that thick and rugged ice mass which is a challenge to icebreakers. In the summer of 2007, a buoy measurement showed 2 metres of bottom melt off a thick ice floe, in a region where first-year ice had reached only 1.5 metres in thickness. Small wonder that all the first-year ice disappeared. The measurements that I have made from British submarines from 1971 to 2007 have shown that the thinning of ice due to global warming has amounted to more than 40 per cent in 36 years, turning the Arctic ice pack from a rugged solid cap over the top of our planet into a thin shell, like an albatross's egg thinned by DDT that will suddenly just crack and break up. Many of us feel that this tipping point has now been reached, and the retreat of arctic ice will now accelerate so that in 20–30 years the summer will be completely ice-free.

The enormous implications of this change for human life and the economics of our society are explored in this book. We must get used to the idea that the top end of our planet in summer has become blue instead of white when viewed from space, and that we in the northern hemisphere, whether it be Europe, Asia or North America, are after all living around a real ocean and not an ice cap.

Peter Wadhams
Professor of Ocean Physics, Department of Applied Mathematics and Theoretical Physics, University of Cambridge
Formerly Director, Scott Polar Research Institute

Preface

Summer comes late in the Arctic. At the back end of August 2008, I came across a newspaper report that both coastal passages across the Arctic Ocean were now free of ice. The news broke from a persistent groundswell of warnings that the polar ice cap was melting. For a journalist like myself, or indeed anyone with a lifelong interest in maritime affairs, it suggested an amazing possibility – that after centuries of cruelly disappointed exploration, the Arctic might soon provide a usable short cut between Europe and China, a North-East or a North-West passage.

It was this thought that triggered the investigation presented in this book. But of course everything turned on whether the scientists were correct in predicting an ice-free arctic summer, and how soon that might occur. The startlingly cold northern winter of 2009–10, for example, has sent them hurrying back to their computer models. Moreover, the possible outcome, while seen by some as an opportunity, was regarded by most of those actually studying the underlying trends – the meteorologists, oceanographers and climate change specialists – as a potential disaster. For them, the Arctic was a dangerous 'amplifier' of global warming.

I need to make clear, therefore, that this is not an exercise, at second or third hand, in predicting the speed of the Arctic's climatic change. It takes the various hotly debated scientific forecasts merely as a starting point from which to discuss the implications – physical, political, economic and military – of a potentially ice-free polar sea. The sociological repercussions for the Arctic's indigenous peoples are also, of course, of immense importance, but largely beyond the scope of this book.

I have sketched in the historical background, not just for the way it elucidates current attitudes, but also for the extraordinary human stories it contains. As for the future, the arctic basin, with its vast hidden resources and contested boundaries, is a cauldron of economic potential and political conflict. What happens there matters to all of us. And scientists struggling to understand its complex, rapidly changing climate fear it may spring more than one nasty surprise.

Meanwhile several things have happened since my main text went to press, serving substantially to reinforce its analysis. This year's seasonal melt – closely matching in extent the record year of 2007 – has emphatically maintained the trend towards ice-free summers. The North-East Passage in particular looked wide open. On southern shipping routes to which the Arctic may one day offer a regular alternative, the trend towards yet bigger container ships continues and there is little sign of a solution to the piracy problem. Most encouraging for hopes of the arctic basin's orderly exploitation was the new agreement between Russia and Norway on division of the Barents Sea's continental shelf – a welcome omen of future co-operation. Most controversial was Exxon's deal with Russia's Rosneft to develop offshore fields in the Kara Sea. Taken together with other companies' plans to drill in such ice-strewn waters, this prompted some to predict a potentially disastrous accident that would make the Deepwater Horizon oil spill look like a minor inconvenience.

David Fairhall
October, 2011

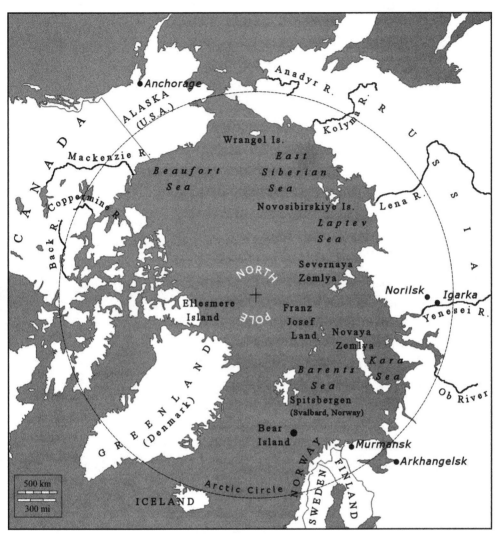

Map 1. A birds-eye view of the Arctic Ocean, looking down on the North Pole and stretching southward in every direction to beyond the Arctic Circle, limit of the midnight sun. Such maps have hitherto been unfamiliar, understandably relegated to the back pages of conventional world atlases – but thanks to global warming, that neglect is set to end.

Introduction

On August 27, 2008, a satellite looking down on the Arctic Ocean observed something possibly unprecedented in human experience. Certainly for the first time in the region's short recorded history, both the fleetingly navigable routes that skirt this frozen sea – the North-West Passage, and the North-East Passage Russians usually refer to as the Northern Sea Route – were ice-free at the same time. For a few weeks that late summer, a ship could circumnavigate the North Pole without being trapped between massive sheets of ice and the bleak shores of northern Siberia or the Canadian archipelago.

At first sight, this rare climatic coincidence might seem to be of no more than academic interest. The North Pole is after all just a theoretical, symbolically significant point in an empty, frozen sea. Its circumnavigation is a largely meaningless exercise. But *crossing* the polar ocean is altogether a different matter. Sailors began probing for a safe way through its many dangers hundreds of years ago, almost as soon as they realised the world was round.

They were searching for a short cut from the Atlantic to the Pacific – a direct route from Europe to China and the East Indies that would save many thousands of miles by comparison with the long haul round Cape Horn or the Cape of Good Hope and across the notoriously stormy southern ocean. Eventually, two partial short cuts were artificially created in the form of the Suez and Panama canals – both of them enormous advances, yet neither eliminating the potential benefits of a polar route.

Many charts use a Mercator projection which grossly distorts the polar regions. And even without that distortion, our egocentric atlases can affect the whole way we envisage the world. A child growing up in Western Europe (I speak for myself) will imagine that to reach Japan by way of the North-East Passage, a ship must turn right at the top of Norway, then sail way across the map before turning down again into the Pacific. Whereas in fact, a straight course past Norway and across the North Pole – using a chart based on an azimuthal projection – leads directly to the Bering Strait between Alaska and Siberia, where the North-East and North-West passages converge.

Distance savings of up to 50 per cent are therefore possible on important trade routes – with commensurate financial benefits – if only some way can be found to overcome the ice barrier. A container ship sailing from Western Europe to Japan, for example, could save several thousand miles by using a polar route instead of the Suez Canal.

But such navigational arithmetic was only one of many reasons why the dramatic ice shrinkage reported in 2007 caused such widespread excitement. An American geological survey had recently alerted us to the fact that something like a quarter of the world's as yet unexploited reserves of natural gas and oil – control of which would be an immense strategic prize – were probably locked up in the arctic basin. The region had also been identified as one of the first places where the effects of global warming were being felt – indeed more than that, it served as an 'amplifier' of those effects.

Cold comfort

The existence of this cold reservoir, insulated by a reflective blanket of ice beneath which deep water is constantly on the move, has a powerful influence on the earth's climatic system. Any major change will certainly affect Europe's weather, both through the atmosphere and, potentially, by disrupting ocean currents like the Gulf Stream. So when scientists already struggling with the threat of global warming saw the polar ice cover rapidly dispersing, the alarm was sounded.

A slow, almost imperceptible melting process had long been detected. Based on this, earlier calculations put the possibility of an ice-free arctic somewhere at the back end of the century. But such predictions were now revised sharply downwards. There was talk of completely open water throughout the summer within a couple of decades. One authoritative prediction even suggested it could happen as soon as 2013!

Meanwhile ship owners hit by rising fuel costs glimpsed the possibility of big potential savings on the long hauls between North-West Europe and the Far East. The Bremen-based shipping group Beluga, for example, quickly began planning to try out the new route.

Lawyers representing the arctic states, having spent decades leisurely processing their clients' conflicting claims to adjacent segments of the polar basin, realised they had better get a move on. Vast offshore reserves of oil, and even more gas, could suddenly become accessible to whoever could establish a legal title. Long-standing disputes about the application of the UN Law of the Sea must quickly be resolved.

Was the sea bed at the North Pole really an extension of the Siberian continental shelf, for instance, and therefore part of Russia's 'Exclusive Economic Zone'. Or was it part of Greenland, and therefore Danish?

In August 2007, a submersible launched from a Russian icebreaker tried to pre-empt the answer by planting its national flag immediately beneath the Pole. An indignant Canadian government dismissed this as an outmoded stunt – having signalled its own ambitions only weeks before by announcing plans for a new fleet of arctic patrol vessels, operating from a new deep-water port.

Introduction

Was Canada meanwhile entitled to claim the North-West Passage as part of its internal waters? Or should it be designated an international strait, as the USA has always maintained – though without ever pushing the argument to a conclusion because until now, it hardly mattered?

What was the modern status of the Spitsbergen (Svalbard) Treaty of 1920, giving Britain, as well as Norway, the USSR and the USA, rights to exploit coal deposits on these remote arctic islands? When the Russians reluctantly signed up to this deal, they had after all only just emerged from revolution.

And what of the strategic naval implications of an ice-free Arctic Ocean? Submariners were certainly keeping a watchful eye on the situation. Whereas diesel-powered submarines like to keep well clear of ice, because they must surface periodically to charge their batteries, nuclear propulsion turned the polar ice cap into a perfect Cold War hiding place from which NATO submarines could keep a listening watch on missile carriers of the Soviet Northern Fleet moving to and from the Atlantic.

Claustrophobia

For Russian admirals, talk of a North-East Passage is calculated to trigger a deep historical neurosis. They have always been conscious of having scattered fleets bottled up in widely separated waters – the Barents Sea, the Baltic, the Black Sea and the Sea of Japan. Way back in 1905, Admiral Rozhestvensky led a squadron of 40 warships 13,000 miles round the world from St. Petersburg at a ponderous eight knots, trailing weed, decks covered in coal, to relieve a Russian squadron trapped in Vladivostok – only to be annihilated by the Japanese at the island of Tsushima. It was this disaster that prompted the construction of icebreakers for an attempt to open up the polar route to the Pacific just before the outbreak of the Great War.

During the Second World War, Russian ports at the western end of the passage – Murmansk and Arkhangelsk – were the destination for allied supply convoys among which German U-boats wreaked terrible havoc. And ironically it was a German vessel, the armed merchant raider *Komet*, disguised under the Russian ensign, which first made strategic military use of the arctic seaway by slipping through to the Pacific during the brief period when Stalin and Hitler were allies.

Now, after years of near dereliction, the post-Communist Russian navy is once again flexing its muscles. And possible conflict over the exploitation of the Arctic's valuable resources – what President Dimitry Medvedev recently referred to as 'our national heritage' – is doubtless a useful justification for naval funding bids; witness the reappearance of formal arctic patrols out of Severomorsk for the first time since 1991.

For Russian sailors, naval or otherwise, opening up what they call the Northern Sea Route has been an historic ambition since the days of Peter the Great, and it was strongly revived under Stalin. In 1932, the Soviet leader established a major government bureaucracy – the Chief Directorate of the Northern Sea Route or Glavsevmorput in the characteristic Russian abbreviation – to pursue it.

The Trans-Siberian Railway had linked Moscow and Vladivostok with a thin ribbon of development since the turn of the century, but vast tracts of Siberia, its timber, minerals and furs, were only accessible by way of the great rivers – the Ob, the Yenesei and the Lena – emptying into the arctic basin. It was imperative for the USSR's planned internal economic development, quite apart from dreams of a complete North-East Passage, to link their estuaries with ice-free Murmansk and Vladivostok.

All this, and the terrible record of the Soviet *gulag*, is part of the historical baggage modern Russians carry with them as they confront the possibility that melting sea ice and the scent of oil will open up their long-secluded arctic seaways to international traffic. For other arctic nations the history is less dramatic, but it still colours public attitudes – towards the indigenous Inuit communities of Canada and Greenland, for example, or the political clash between former Alaskan governor Sarah Palin's policy of 'Drill, baby, drill' and the desire to preserve inviolate, stretches of the northern wilderness.

'I may be some time'

For British people polar exploration, Arctic and Antarctic, is both part of our maritime tradition and a prolific source of public heroes – Parry, Franklin, Shackleton, Scott and that tragic exponent of English understatement Captain Oates. Even the 14-year-old Nelson was once there, trying to kill a polar bear.

While Russian, Dutch and Scandinavian explorers painfully pieced together the eastern arctic route, British efforts were concentrated on the elusive North-West Passage across what is now Canada. Here, ice was just as much of a problem, but there was the added difficulty of finding a way through an unmapped archipelago. There were dozens of ice-strewn bays, sounds and inlets, where ships could be trapped through a sunless winter – often literally dead ends.

The British navy's search reached a grim climax in 1845, when Sir John Franklin's expedition – two ships carrying more than a hundred men – effectively disappeared in this maze of islands. Over the next ten years more than 40 search parties were sent out, many of them funded by Franklin's loyal wife Jane, who became a Victorian celebrity in the process. Her passionate campaigning continued long after he must be presumed dead, and the search did eventually find a few sad relics.

The key to progress along the North-East Passage was the development of the icebreaker, a powerful, specially adapted ship that can master the ice rather than submit to it. At the end of the nineteenth century, the Imperial Russian Navy led the way with the steam-powered *Yermak*, built by Armstrong Whitworth at Newcastle-upon-Tyne. In the 1930s, another icebreaker, the *Sibiryakov*, helped obtain Stalin's crucial backing for the creation of a powerful new bureaucracy to promote the arctic route by completing the first east–west transit in a single season.

A different league

But it was the appearance of the nuclear-powered *Lenin* in 1959 which put the Russians in a different league where icebreakers are concerned. Just as the British applied their nuclear bomb technology to civilian electricity generation in the so-called 'Atoms for Peace' programme (a spin doctor's classic), so the Russians used their experience of submarine propulsion to build immensely powerful vessels which can go for years without refuelling – the world's only nuclear merchant ships.

Icebreakers are strange, almost freakish vessels. Structurally complicated and therefore extremely expensive to build, they have numerous special features to do a special job – starting with a characteristically sloping, sawn-off bow so as to ride up on the ice and break it with the ship's weight. Add to this the complexities of a nuclear power plant – driving turbines supplying electricity to drive multiple propellers – and it is not surprising that the Russians have had serious problems with their icebreaking fleet as well as many spectacular successes. There have been leaks and accidents, some alarming pollution, and since the collapse of the Soviet Union, wider fears about the security of other nuclear plant and radioactive materials. Anxious foreign governments, especially in neighbouring Norway, have provided technical and financial support as a matter of self-interest as much as philanthropy.

The nuclear fleet has never been directly profitable (not even, I would guess, by the arbitrary standards of Soviet accounting).

Chaotic privatisation of large parts of the Russian shipping operation therefore led to recurrent arguments about pricing, funding and ultimately the fleet's costly replacement. In 2008, these culminated in a strategic Kremlin decision to transfer management of the ships from the privatised Murmansk Shipping Company to the state nuclear agency Rosatom.

Known unknowns

How this plays out in the context of the Northern Sea Route's redevelopment remains to be seen. Moscow may well be waiting, like everyone else, to see how fast the Arctic melts. But for the time being, exploitation of the oil, gas, metals and timber on which Russia's future economic strength substantially depends, cannot be achieved without icebreaker support and/or the introduction of more specialised icebreaking cargo vessels.

The head of Rosatom has promised a new generation of nuclear ships. The mighty Norilsk mining combine, probably the world's leading producer of nickel and palladium, and already exporting year-round from the Siberian river port of Dudinka, has meanwhile pioneered an extraordinary new kind of icebreaking container ship, a so-called 'double-acting' ship which ploughs through ice by swivelling its propellers through 180 degrees and travelling backwards.

The extent to which ordinary ships must be ice-strengthened to take advantage of the polar routes – with or without icebreaker assistance – is one of the first questions interested foreign owners will now ask. Russian vessels intended for these waters are routinely hardened and adapted in varying degrees, depending on how far east from Murmansk they need to penetrate, and whether they have to face harsh winter conditions. Newcomers will have to confront these same problems and, wherever possible, will prefer to be independent of expensive icebreakers.

When the Soviet economy crashed in the 1990s, the Northern Sea Route crashed with it. But a rapid revival, at least at the western end, is now being driven by oil and gas development. Within five years, oil traffic across the Barents Sea is expected to reach 20 million tonnes a year as fresh fields are exploited to meet European and US demand. Sovcomflot, the state-owned shipping company, has announced plans for a million tons of new oil and gas tankers, and the giant Gazprom has confirmed that as a producer, it needs ships to reduce its dependence on pipelines. Meanwhile the traditional trade in timber seems to be reviving, requiring a new fleet of specialised freighters.

The geographical hub of all this activity is Murmansk (with its associated naval port of Severomorsk), kept clear of ice even in winter by the warm Atlantic water of the Gulf Stream. It is the home of the privatised Murmansk Shipping Company, which has been at the forefront of arctic maritime development since it was established under the Soviet regime in 1939 – except that it has now, for good or ill, lost control of the nuclear icebreaking fleet.

An invitation from Gorbachev

In 1987, Mikhail Gorbachev chose Murmansk to make a major policy speech in which he suggested that the Northern Sea Route – which in practice had hitherto been regarded as a Soviet preserve – should be offered to the international shipping community. The North Pole, he said, should become 'a pole of peace'. And four years later, the route was indeed formally declared open on a non-discriminatory basis.

Nothing happened. Hardly anyone took up the invitation. But 20 years on, in a different physical and economic environment, the Russian maritime authorities are renewing their efforts to provide the necessary infrastructure – charts, navigational aids, weather and ice reports, marine communications and so on.

That is at the technical level. Before foreign ship owners are prepared to invest serious money, they will also be looking for signs of political and financial commitment from Moscow. And of course they will be watching the satellite ice maps to see if the spectacular shrinkage of the past few years is maintained, and how the North-East Passage is likely to compare with a North-West Passage under more predictable Canadian, rather than Russian, control.

Other things being equal – which unfortunately they are not – the potential savings in distance look immensely attractive. The arctic route between Holland and Japan, for instance, is about 4,000 nautical miles shorter than a voyage south-about through the Suez Canal – a bigger absolute reduction than was achieved on some routes by the opening of the canal in 1869, before which ships bound for India and beyond had to make the long haul round southern Africa. If it were physically possible, one of those vast container ships leaving Felixstowe for Shanghai could save about 2,000 miles by going across the arctic. Some passages could be shortened by up to a fortnight.

There are clearly big incentives here. Fuel is a large component of ship-operating costs, so cutting the total burn by a third would be a massive bonus. But against that must be set the costs and uncertainties of operating in a new and often hostile maritime environment – cold, foggy and often stormy. Whatever the eventual outcome, the Arctic is not going to melt overnight and for the foreseeable future is not expected to remain unfrozen for more than a few months at a time. Difficult ice, accumulated over several years, will persist in some areas. Even if there is soon open water through the summer season, the winter cover will return each year, and where there *is* less ice, it may be dangerously on the move.

Nevertheless, given the probability of a relentlessly warming climate, a major realignment of the world's maritime trade routes – the biggest since Ferdinand de Lesseps dug his wonderful desert canal between the Red Sea and the Mediterranean a century and a half ago – is at least a possibility. That was the starting point for this book – a sailor's reaction, if you like, to the news that the two arctic passages were simultaneously open. Yet the moment one begins to follow through the implications of this in one area, the conclusions fall foul of uncertainties in another. A narrow maritime perspective widens to take in the more worrying implications of climatic change – the arctic 'amplifier' at work.

If the ice melts, so does the frozen shoreline – the tundra – with the potential to release vast quantities of methane, far more effective as a 'greenhouse' gas than the carbon dioxide we are usually warned about. Greenland's icy shroud will also shrink, along with its surrounding pack ice, perhaps disturbing deep ocean currents that could trigger climatic change far more abrupt than the effects of gradual atmospheric warming – by weakening the Gulf Stream which warms North-West Europe. Our pursuit of a North-West Passage would seem trivial by comparison.

Come what may, over the next couple of decades the arctic basin is set to become economically, and hence politically, much more important. So many of the world's limited sources of oil, gas and minerals are still to be found there – mainly to the Russians' benefit even if the retreating ice helps other arctic nations grab a share. Whatever strategic decision the Kremlin takes about rebuilding Russia's military strength, therefore, it will acquire powerful economic leverage in its dealings with neighbours, partners and competitors. How that plays out depends on political events that have nothing to do with oil or ice.

Unknown unknowns

I referred earlier to the many obvious questions about the Arctic's future to which the answer is uncertain, some of which I have touched upon briefly in this introduction. But talking to climatologists in particular brings it home that rapid change, with so many variables

INTRODUCTION

at work, may well spring complete surprises about which we cannot even frame the appropriate questions. The following pages therefore contain few predictions – at least not from me. They merely attempt to sketch in the factors that will influence events as they occur – the history, the climate, the politics, the economics and the technology – because one way or the other, they are going to be of great importance to us all.

For me, the 'Cold Front' of the book's title is a meteorological analogy – turbulence, perhaps developing into a gale, followed by showers and bright periods. Outlook uncertain.

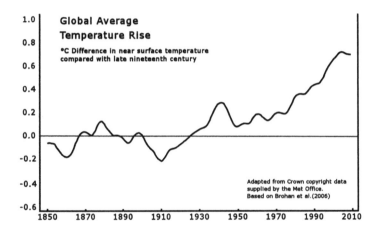

THE ARCTIC ARENA

Arctic – from Greek arktos: *bear*

Mary Shelley's novel *Frankenstein* begins and ends on the frozen waters of the Arctic Ocean. Her narrator is an explorer, captain of a vessel trapped in the ice, when both the story's murderous monster and his tormented creator just happen to come by on sledges, one pursued by the other.

A ludicrously implausible encounter, but of course the author was less concerned with technical authenticity than with atmosphere, the sense of gothic horror she wanted to create. And this extreme environment, its deceptive beauty disguising constant danger from ice or cold – and in 1817 still deeply mysterious – suited her purpose well.

The term 'arctic' comes from the Greek word for bear: not the polar bear so often used as its symbol, but the constellation of the Great Bear, always visible in the northern sky. Two of the stars which make up that familiar saucepan shape point to the Pole Star – a double star, though not recognised as such until late in the eighteenth century when William Herschel's new telescope, set up in his back garden in Bath, separated its two components.

Double or not, it had long been invaluable to navigators because it appears in the sky almost exactly above the terrestrial North Pole. For many centuries it really was a guiding star, whereas the

Magnetic Pole, to which a compass needle points, was not only a long distance from the true pole – somewhere in the Canadian arctic archipelago – it also moved about.

Sailors using traditional methods still make allowance for this 'variation' between true and magnetic North, and in arctic waters the difference may be enormous. The early explorers searching for a North-West Passage to the Pacific found their compasses almost useless at times, so that finding the exact location of the Magnetic Pole became a major preoccupation. Indeed Mary Shelley's fictional explorer was also searching, among other things, for 'the wondrous power that attracts the needle'.

The neatest definition of the arctic region is this: that part of the globe enclosed by the parallel of latitude 66 degrees, 33 minutes N – the Arctic Circle. It marks the southern limit of the 'midnight sun' at the summer solstice, and the northernmost point at which the midwinter sun briefly appears. By this measure, there are eight arctic states, all members of the Arctic Council, a body formed in 1996 which also represents a number of indigenous groups and already plays an important part in preparing for the consequences of climate change.

The Arctic Ocean is vast, empty and deep. It fills a geological basin roughly 1,000 by 2,000 miles in extent, with access from the Greenland and Barents seas at the European end and a narrow entrance from the Pacific – the Bering Strait – at the other. In more than one sense, it is the opposite of the Antarctic continent, where a frozen land mass is surrounded by the largely ice-free southern ocean. Here in the North, five of the arctic nations – Canada, Denmark (through Greenland), Norway, Russia and the USA – look out across an icy reservoir that both unites and divides them, nowadays watching with increasing wariness lest one of their neighbours should grab more than their share of its resources.

Thickness matters

For up to eight months of a long winter, and progressively from October to March, the sea freezes – starting at a temperature of minus 2 degrees C. It forms almost continuous pack ice averaging 4–5 metres (13–16ft) in thickness, including thin first-year ice and pressure ridges of up to 20 metres or more. (At the North Pole, which paradoxically is less cold than some of the surrounding land mass, the mean is 3–4 metres.). So the actual thickness varies widely. There are always some patches of open water. And of course this matters a great deal, both to vessels forcing a way through and to anxious climate scientists trying to estimate the volume of ice remaining as the arctic meltdown accelerates.

Whereas the changing area of the ice can easily be monitored from satellites, measuring thickness from space is a more complicated business, for which the European Space Agency's newly launched CryoSat-2 has specifically been equipped. Surveys conducted so far suggest that the ice is rapidly thinning, but to confirm that, scientists are turning to any source they can find – scattered reports from icebreakers, special projects like the Catlin mobile radar survey from northern Canada, aerial magnetic surveys and, most usefully, underwater measurements by upward-looking sonar on submarines – unlikely recruits to the study of climate change while continuing to rehearse their secretive wartime role.

From April to September, as the ice gradually melts, large stretches of the ocean fringe become clear – or almost clear – since drifting patches of ice remain, along with 'bergy bits' and the especially dangerous 'growlers', barely visible above the surface. In past summer seasons, the central pack has eventually shrunk to perhaps half its winter size, but continued to choke the northern part of the Canadian archipelago and cling to the east coast of Greenland. More importantly – at least from a Russian perspective – obstinate tongues of ice still reached out towards the Siberian shore, especially where the islands of Severnaya Zemlya ('Northern Land') mark the entrance to the Laptev Sea, impeding coastal shipping on which some remote communities are almost totally dependent.

Not many people positively choose to inhabit this bleak environment – though the dwindling indigenous groups who still herd reindeer or hunt whales may protest otherwise. It has a dramatic, monochromatic intensity which evidently attracted many European explorers, just as others were fascinated by the empty sands of Arabia, but living there permanently is another matter.

The arctic shorelines consist predominantly of tundra (from a Lapland word meaning 'treeless'), so cold or dark for much of the year that only small vegetation can survive – heath, grasses, moss, lichen and a surprising number of hardy flowers such as purple saxifrage and the arctic poppy. Apart from migrating birds

1. An endangered species.

and reindeer (caribou), indigenous animals consist either of highly adapted species like the polar bear and the arctic fox – both of which have been encountered far from land near the North Pole – or those such as whales and seals which find refuge in the sea.

Most specialised among the whales are the white beluga and the narwhal, with its wonderful spiral tusk. The first Europeans to encounter the narwhal thought this implausible creature might be a magical 'sea unicorn'. Queen Elizabeth I was certainly delighted to be presented with a tusk, and generations of arctic travellers have brought one home as a souvenir, perhaps to be mounted above the bar of their local inn alongside the boar's head and the antlers.

The giant among arctic whales is the bowhead, which grows to more than 20 metres (65 ft) in length and is believed to live for up to 150–200 years. Unfortunately for its own survival, it belongs to the family of right whales – identified by the early whalers as the 'right' kind to catch because it often swam slowly along the shore, floated when it was killed and produced baleen as well as blubber. It migrates short distances with the seasons but stays within arctic waters.

Coming in from the cold

When it comes to migration, it is of course the birds which lead the way. Indeed they provide a permanent seasonal link between this remote region and the comfortably temperate environment in which most Europeans live.

The brent geese which descend on the Essex marshes near my home every autumn have flown in from the Taimyr Peninsula, the northernmost point of the Siberian mainland at the crux of the North-East Passage. By counting the youngsters among them, my expert birdwatching friends can even tell what sort of summer the geese have had in the far north. If there was a plentiful supply of lemmings (another characteristic arctic animal), the arctic foxes will not have robbed so many birds of their eggs.

In those high latitudes, a summer temperature of more than about 10 degrees C rates as warm, while in winter, temperatures drop to minus 30–40 degrees C – and far colder than that in the depths of the Siberian 'taiga'. In many areas the subsoil remains permanently frozen through the summer, leaving surface water to collect in boggy, mosquito-ridden pools. Precipitation, either as snow or rain, is light. Climatically, this is a cold desert. Growth and decay are symmetrically slow – which, importantly in the present context, makes the tundra especially vulnerable to industrial pollution. Nowhere is an oil spill more likely to linger.

In spite of the underlying cold, the long days of summer sunshine have sometimes been sufficient to open up coastal sea passages between the Atlantic and the Pacific, or at any rate large stretches of them, that were just about navigable. They might lead north-west through the Canadian archipelago or north-east along the Siberian coast, but there was always some ice around. So until now these routes have only been safely available to specially ice-strengthened ships or with the assistance of icebreakers to clear a path for the other vessels. The major exception was the Barents Sea, that western corner of the Arctic warmed by the Atlantic Gulf Stream. Both the North Cape of Norway and the nearby Russian port of Murmansk, with its associated naval bases, were always ice-free.

But the accelerated melting of the past few years suggests that transit routes right across the Arctic may soon be completely clear during the summer months. The North-East Passage, known to the Russians as the Northern Sea Route, opened briefly in 2005 – only to close again. Two years later, the Canadian North-West Passage was fully navigable for the first time in many years. Then in 2008 – on August 27 to be precise – a NASA satellite showed both passages simultaneously open for the first time in recent recorded history, and possibly for much, much longer than that.

An alarming opportunity

This rare coincidence came as a surprise to scientists and seafarers alike. For most of the former, who saw changes in the arctic climate as a harbinger of dangerous, potentially catastrophic global warming, it was a matter of serious alarm. 'The Arctic is screaming', as one excited American observer put it. An ice-free polar region implied melting glaciers, rising sea levels and severe meteorological disturbance – possibly including the extinction of the Gulf Stream which gives North-west Europe its temperate climate – and the release of vast quantities of methane trapped by the permafrost that would drastically exacerbate the greenhouse effect.

However, for those struggling to work within the Arctic in its present frozen state – supplying remote communities, exporting timber and minerals down Siberian rivers or drilling for offshore oil and gas – the thaw promised to make things a lot easier. What is more, shipping companies currently sending cargoes right round the world by way of the Suez or Panama canals were offered the prospect of a polar short cut saving thousands of miles on a voyage, say, from Rotterdam to Yokohama, or northern Alaska to New York. Ice-bound channels that had defied mariners' best efforts for 500 years might suddenly be plain sailing.

There is a paradox here, however, at least for those of us unused to arctic conditions. Polar bears are not the only ones to find a solid covering of winter ice useful – in their case to venture out on to the frozen sea in search of seals. Ashore, given the scarcity of decent roads and bridges, it is often much easier for human beings to move around when ground and rivers are frozen than when the summer thaw turns them into treacherous bogs or torrents. Even in coastal areas, a snowmobile (a sort of tracked scooter) will reach places that a boat cannot. In northern Canada, winter conveniently transforms the Mackenzie River delta into an 'ice road' down which vast trucks trundle smoothly – if somewhat perilously – to supply outlying communities, drilling rigs or military installations. Some Siberian rivers are used in the same way.

Nevertheless, at sea ice is both inconvenient and dangerous. And it takes many forms. The polar pack ice is not the simple, uniform thing southerners like myself might imagine. Just as British or Dutch seamen talk of creeks and channels, swatchways and deeps, their Russian and Canadian counterparts distinguish between many different kinds of ice – young, old, landfast, layered or rotten. Churned up by the sea, it may become pancake ice, or form vast floes, weighing literally millions of tons, which grind together to form ridges, high and deep. Offshore, there is always some movement. A crew member aboard the *Sedov*, a Russian vessel trapped in the arctic pack for more than two years at the beginning of the Second World War and in constant danger of being crushed, recalled the dread sound of ice on the move: the sinister hiss of slabs climbing

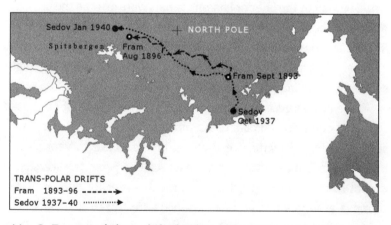

Map 2. Two vessels have drifted right across the frozen Arctic Ocean - the *Fram* in a deliberate experiment, the *Sedov* accidentally beset.

over one another, a rattle like machine gun fire, the rumble and roar as floes shattered under the enormous pressure.

The basic rule is that the colder, older and thicker the ice, the more difficult it will be to negotiate. Worst of all is really old ice crumpled into a rock-like mass over four or five years. Even a powerful icebreaker may then have to resort to repeatedly backing and filling to smash a way through.

Beneath this frozen cover, the whole ocean is churning away under the influence of wind, temperature and salinity, moving vast masses of ice around in the process. The explorer Nansen's vessel, the *Fram*, deliberately sailed into the ice in the 1890s just to see what happened, and the unfortunate *Sedov*, unintentionally trapped 40 years later, showed that after entering the pack from the Siberian shore they would slowly drift westwards. Wreckage of the American De Long's little *Jeannette*, crushed after entering the Bering Strait in 1879 in search of an imaginary polar land mass, had done the same, turning up five years later in Greenland – indeed it was this that prompted the Norwegian explorer's triumphant experiment.

Fridtjof Nansen was extraordinary not just for his almost superhuman fortitude – something he shared with other arctic pioneers. His personal adventures – sledging across the Greenland ice cap and drifting across the polar sea – contributed to scientific study that is still being followed through today in an effort to unravel the complexities of global warming. And although he was prepared to face years of arctic isolation, he never lost a wider human perspective. 'Oh how tired I am of thy cold beauty', he wrote in his journal of the *Fram* voyage. 'Life is more than cold truth, and we live but once.'

It was impatience for life as well as a chance of sledging to the exact North Pole that prompted him to leave the *Fram*, with just one companion, Lieutenant Frederik Johansen, while the ship was still trapped. After reaching a record 86 degrees N, about 200 miles from the Pole, they retreated to Franz Josef Land (discovered not long before by an Austrian expedition whose leaders named it, inevitably, after their emperor). There the Norwegians spent the dark winter in a hut improvised from stones, driftwood and walrus hides before heading south again, their sledges lashed across a pair of kayaks to form a sort of catamaran.

The barking dog

One morning, still many hundreds of miles from the mainland, let alone safety, Nansen heard a dog bark. He knew it must be accompanied by a human being, and the sound led him across the ice to an extraordinary encounter comparable only to the legendary meeting between Livingstone and Stanley on the shores of Lake Tanganyika. The other human turned out to be an Englishman, Frederick Jackson, out there to do a spot of exploring sponsored by Lord Northcliffe's newspaper.

Jackson could not quite match Stanley's 'Dr Livingstone, I presume.' Understandably confused by the Norwegian's blackened face and grease-laden clothes, he could only manage: 'Aren't you Nansen?... By Jove!', followed by a nice touch of courteous English understatement – 'You have made a good trip of it, and I'm most awfully glad to be the first person to congratulate you on your return.' (Perhaps Mary Shelley's arctic scene was not so far-fetched after all?)

The ice-laden current which carried Nansen's small wooden vessel is nowadays identified as the Transpolar Drift, and the channel between Greenland and Spitsbergen through which it spews large quantities of ice into the Norwegian Sea is sometimes referred to as the Fram Strait. The other large-scale movement within the arctic pack is the Beaufort Gyre, a circular mass rotating clockwise North of Alaska; apparently it takes about four years to complete one rotation.

Measurements from the *Fram* and the *Sedov* showed that below the ice of the Transpolar Drift a layer of relatively warm water – the tail end of the Atlantic Gulf Stream – reaches deep into the arctic basin from the opposite direction. And beneath that is a third, heavily saline layer (sea water releases its salt as it turns to ice), balancing the system by moving back towards the Atlantic. Some warm Pacific water also enters through the Bering Strait, to skirt the Alaskan and Canadian shores.

Although the ocean is pierced in several places by mountainous outcrops such as Spitsbergen and Franz Josef Land, there are large areas of seriously deep water. The Russian submersible *Mir* had to descend to more than 4,000 metres in 2007 to plant its national flag at the North Pole, where, incidentally, for all its symbolic importance, the seabed proved remarkably uninteresting – just yellowish gravel. But not far from the Pole is a prominent underwater feature geologists are extremely keen to explore, the Lomonosov Ridge. This stretches right across the basin from near Greenland to the Novosibirskiye (New Siberian) islands, and the question is whether it amounts technically to an extension of the continental shelf on either side – a matter of great consequence under the international law of the sea.

To clear open water for the flag-carrying submersible's mother ship, the Russians had to use one of their big nuclear powered icebreakers, the *Rossia*. Whether we shall ever see conventional ships operating at such latitudes without icebreaker support no one can confidently say. No scientist yet pretends to comprehend the full complexity of arctic climate change. The data are incomplete, the mechanisms not yet fully understood and the records – many derived from satellites and submarines that were simply not around 50 years ago – perhaps misleadingly short. But recent events nevertheless provide powerful evidence that the Arctic will soon live up to its reputation as the world's 'weather kitchen' by producing global warming's first unmistakably momentous event.

Tipping point

The polar basin has also been described as 'the amplifier of global warming'. While it remained shrouded in ice, most of the sun's rays were reflected back into space, with little direct melting effect and little chance to warm the underlying water. Any clear patches that appeared in the summer were quickly refrozen as winter returned to restore an icy equilibrium.

Nonetheless, average temperatures across the region have slowly been rising over the past 50 years. Year on year, this change may have been imperceptible to anyone but an Inuit hunter whose livelihood depends on the seasons, but the Arctic Council nations were concerned enough to organise an exhaustive *Arctic Climate Impact Assessment,* published in November 2004, which found that the temperature rise was almost twice as fast as in the rest of the world.

And as you would expect, warmer air has been accompanied by shrinking pack ice. Records recently analysed by US scientists indicate a reduction in summer ice cover of about 10 per cent a decade since 1979, when accurate satellite images became available.

Crucially, however, the linkage between these two phenomena is more than direct. The warming effect is amplified by a feedback process. Warm air melts the ice from above, uncovering some water in the late summer. And the larger the area of water, the more the sun's warmth can be absorbed by it and subsequently dispersed, instead of being reflected from the ice – thereby strengthening the melting effect in two ways, from the sea below and the air above. Once triggered, the process is likely to accelerate, rapidly depleting the ice unless some new factor, such as changing weather patterns, comes into play.

That acceleration is just what some scientists now think is happening – witness the record shrinkage of 2002 and 2005, followed by the startling summer melt of 2007, leaving both arctic passages briefly navigable in 2008 for the first time since comprehensive records began (perhaps even for more than 100,000 years, since the last interglacial period). This news rocked the climate change community. Relatively simple computer models extrapolating from earlier trends had hitherto predicted a completely ice-free summer some time in the second half of the twenty-first century – maybe by around 2070. But as each September's satellite images appeared and new factors were fed into the calculations – such as less thick, multiyear ice and more warm water flowing in through the Bering Strait – the published forecasts leapt forward, a decade at a time.

The most aggressive prediction so far emerged from a research team at the Naval Postgraduate School in California, suggesting a virtually ice-free summer perhaps as soon as 2013. Professor Peter Wadhams, a hands-on ocean physicist at Cambridge University who has spent a lot of time in Royal Navy submarines measuring the ice from beneath the polar ice cap, agreed that whatever the exact date, it was not far away – the pack ice was simply collapsing.

By 2008, we seemed to be approaching what scientists call a 'tipping point', the point at which a slow process of climate change suddenly develops a runaway speed. If that does happen in the Arctic, other changes will inevitably follow. There will be some painful, even disastrous, consequences. There will also be new opportunities, especially for those nations bordering arctic waters.

Frozen Assets

The Svalbard poppy, blooming suddenly on Spitsbergen as the summer sun unfreezes the topsoil, is a rare variety capable of producing white flowers. It is believed to have survived the last ice age. In spirit, if not in scientific fact, it forms a link with a time when this glacial archipelago was covered with warm, rich vegetation whose remains are now mined as coal by Norwegians and Russians.

Coal is a valued commodity in this part of the world. And back in the early years of the twentieth century, when mining first began on Spitsbergen, it was doubly so because ships used it to raise steam. In more recent decades – and especially during the Cold War – the fact that mining continued in this harsh environment had as much to do with the archipelago's strategic location in the Barents Sea as with any commercial profit. Nevertheless, Norway's modern, state-owned Svea mine exports several million tons of high-quality coal a year. At dilapidated Barentsburg, the remaining Russian mine is almost exhausted and, at the time of writing, was out of action altogether following a fire. There was talk, however, of reopening an older mine at Grumant – a measure, if it happens, of the value Moscow places on maintaining a continued Russian presence.

The other fossil fuels found in the arctic basin – petroleum and natural gas – are nowadays of immense economic importance, but the modern oil industry is only 150 years old. For centuries before that, the arctic communities were exploiting natural resources for which there was already a big demand – animal furs, timber, whales and fish.

Russian company

In the summer of 1553, England's merchant adventurers sent out three small ships to search for a North-East Passage to China and the Indies. Only one vessel survived, getting no further than Arkhangelsk in the White Sea. But there its crew did make contact with Russian hunters who were interested in trading. The English sailors even braved a long inland journey – a round trip of about 1,500 miles – to pay their respects to Ivan the Terrible in Moscow. The reward of their enterprise was the establishment of the Russia Company, whose vessels sailed north each summer to load cargoes of furs and timber. Over the next few hundred years both became staples of Russia's trade with Western Europe, helping to break down her political and cultural isolation.

Russia has a vast amount of timber. The largest segment of that great swathe of coniferous forests which borders the tundra right round the arctic basin – the taiga – lies within its borders, about a fifth of the world's total forested area. But with few decent roads, and until the beginning of the twentieth century not even the Trans-Siberian Railway, exploiting its export potential has never been physically easy except where logs can be floated down rivers. And most of those eventually flow north.

As far as the northern sea routes are concerned, the major port of Arkhangelsk in the west, though hampered by winter ice, has always provided at least one good outlet for the Russian industry. Igarka, much further east and a remarkable 400 miles up the River Yenesei, was established in 1929 specifically as a sawmill and timber export terminal. It became something of a boom town, yet it is deeply frozen for much of the year, as is the Kara Sea – the 'ice cellar' – into which its turbulent river empties. Siberian timber exports, having declined, now seem set for a revival, and a genuinely ice-free summer would obviously transform this situation.

Animal furs, on the other hand, did not pose the same transport problems even back in the sixteenth century. Russians welcomed the trade, and the access to Western products it provided. Here was one of the few rewards for the hardship later endured in opening up Siberia, where arctic fox, ermine (a posh name for stoat in its winter colouring) and the luxuriant, highly prized sable could be hunted.

As North America was colonised by Europeans, it too provided a plentiful supply of furs, although the mainstay of that trade, the beaver – used especially for hats – is not strictly an arctic animal. Only in the twentieth century did fur clothing rapidly become less useful and then less fashionable. Even so, tens of thousands of arctic foxes are still trapped, shot or otherwise slaughtered each year across Russia, Canada and Alaska for their superbly insulated pelts.

Lamp oil and corsets

A quite different group of animals also lured Europe's early merchant seafarers north in search of profit – the whales. In 1607, the Russia Company hired Henry Hudson to make another attempt at a polar crossing, and although he failed in that, his report of Spitsbergen's western coastline teeming with whales, seals and walruses prompted a booming new industry.

In a world where petroleum products and plastics were still unknown, whales were immensely valuable and mercilessly hunted. Their blubber-insulated bodies provided lamp fuel, lubricants, soap and cosmetic oils. 'Right' whales like the bowhead also filtered their watery diet through long strips of baleen, or whalebone, a strong but flexible material ideal for stiffening ladies' corsets.

In spite of the terrible hardships sometimes involved, the English and the Dutch rushed to exploit this new resource, using expertise provided by Basque whalers from the Bay of Biscay ('harpoon' is a Basque word). The shore-based industry on Spitsbergen – based in the Dutchmen's case at the aptly named Smeerenburg (Blubbertown) – lasted only a few decades before coastal stocks were exhausted and whaling fleets took to the open sea in relentless pursuit of their gigantic quarry. In this form, the arctic industry remained active – with many Americans later joining the hunt from Nantucket or San Francisco – until around the end of the nineteenth century. Though petroleum had by then supplanted blubber, rigorous corsetry was much in fashion, so it was still worth chasing down the dwindling whale populations for their baleen.

As with whales, so with fish. Moral arguments for saving the whale aside, these are both scarce, finite resources. They generate fierce human competition. Those exploiting them have rarely been able to agree amicably on how the catch should be shared, let alone when to stop so as to make the operation sustainable.

Cod wars

As a young national service conscript, I caught my first sight of the arctic ice pack from the deck of a fishery-protection minesweeper somewhere off Greenland (our skipper had stopped alongside the ice for a bit of gunnery practice), during the first of the so-called 'cod wars' between Iceland and the UK. We lost that war – and the next. British deep-sea trawler fleets were driven from their traditional grounds, forced either to head straight for the scrapyard or to try their luck in still more distant waters. Years later a Hull trawler skipper told me of a trip he had made way north into the Barents Sea.

Searching for fish – probably haddock – he encountered a strange convoy of Russian freighters heading for the Atlantic under naval escort. They had an odd deck cargo, *so* odd it was worth a radio report – probably the first Western intelligence, as it turned out, of the incipient Cuban missile crisis.

While Alaska is famous for its giant king crabs, the characteristic fish of these polar seas is the arctic cod, a smaller, uglier relative of the Atlantic, 'fish-and-chips' cod. The arctic species actually likes extremely cold water. It can be found lurking under the ice, further north than any other fish, a favourite food for narwhals and white beluga whales. It is also caught for human consumption, especially by the Russians and the Norwegians (the Canadians have not bothered with it in the past, although it occurs right across their northern archipelago).

The Barents Sea cod fishery has for many years been managed by Norway and Russia through a joint fisheries commission. It seems to have worked well to begin with, but over the past decade, familiar arguments about whether the stock is being overfished – and if so who is guilty of the overfishing – have surfaced with a vengeance. And as in other contexts, the two neighbours' differences have been accentuated by the underlying tension over the status of Spitsbergen and a long-standing argument over the boundaries of their offshore jurisdiction. Not a good omen for wider disputes among the arctic nations as melting ice and advancing technology free other, economically much more important resources.

Although the cod fishery is vital to the remote coastal communities that depend on it, its management is of little direct consequence to the wider world. By contrast, there is intense concern about who gains control of arctic oil and gas. Hydrocarbon reserves are finite; demand for them seems insatiable; here are vast new unexplored reserves – albeit extremely difficult to extract.

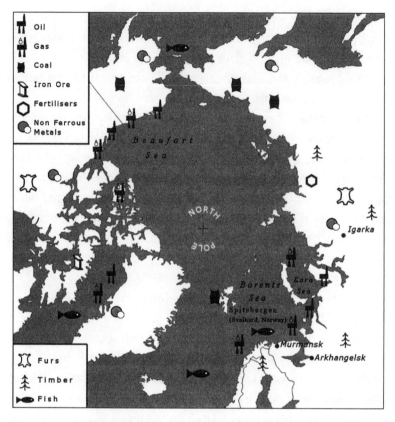

Map 3. The Arctic's Natural Resources.

Striking it rich

In 2000, the US Geological Survey's *World Energy Assessment* suggested that a quarter of the world's undiscovered oil and gas reserves probably lay under the Arctic. In 2008 – by which time more geological data were available and rapid melting of the arctic ice had given its analysis much sharper significance – a greatly refined assessment was published as the *Circum-Arctic Resource Appraisal (CARA)*. So-called unconventional resources like tar sands and gas hydrates were ignored, but any likely oil or natural gas field within

2. The Arctic holds perhaps a quarter of the world's remaining oil and gas.

the Arctic Circle considered recoverable by existing technology – even offshore in deep, ice-covered water – was plotted.

The results were to say the least impressive. The hydrocarbons waiting out there to be tapped by the drilling rigs probably amount to 90 billion barrels of oil (13 per cent of the world's total undiscovered reserves, or enough to keep us going for about three years at present consumption rates), plus 1,700 trillion cubic feet of natural gas (a much bigger 30 per cent of undiscovered reserves) and 44 billion barrels of natural gas liquids (20 per cent). All this in addition to more than 400 oil and gas fields that have already been found within the Arctic Circle, in Alaska, Canada and Russia – about a tenth of all *known* reserves.

Two striking patterns emerge from the American geologists' work. First, more than 80 per cent of the potential new reserves are offshore – but not predominantly in the deep central basin where the Russians planted their titanium flag. Mostly they lie under the surrounding continental shelves where recovery is feasible, if expensive. Second, these resources are not evenly distributed. Most of the gas is in Russia or definitely within its offshore jurisdiction in the Barents and Kara seas. About a third of the oil is likely to be found on the arctic shores of Alaska. Relatively smaller – though certainly not insignificant – amounts of both oil and gas are spread across the Canadian archipelago, off Greenland and Norway (which is already extracting gas from its pioneering Snohvit field in the Barents Sea, for export by tanker).

So while all the coastal nations will have some access to this hidden wealth, Russia's already strong position (especially where gas is concerned) and the USA's stake in Alaskan oil, are likely to be reinforced. Greenland could be an important newcomer, further complicating its strained relations with the central government in Copenhagen.

How far and how fast any of these players exploit their advantage is another matter. It can take ten years to start pumping from a new offshore field even in benign conditions, and arctic conditions are far from benign. For much of the year, engineers have to contend with freezing temperatures, darkness, storms and the effects – still not fully understood – of moving pack ice grinding against floating rigs. If the Gulf of Mexico has problems, future operations in this more hostile environment can hardly be risk-free.

For the Russian petroleum industry there are added problems – its shortage of investment capital, the fact that it has to buy in much of the relevant offshore technology, and Moscow's determination not to lose political or financial control by simply handing the job over to Western oil companies. The way development of the big new Shtokman gas field off Murmansk is being handled reflects all this.

Gas was first discovered there way back in 1988, but quite apart from the physical difficulties of extracting it, the economic chaos which followed the collapse of the Soviet Union meant nothing much useful happened until the early years of this century. By this time Putin's drastic reforms had taken effect and Russia's giant state-controlled Gazprom company was in charge, its strategy shrewdly designed to enlist essential foreign technical and financial support while retaining maximum control.

A state within a state

Gazprom (an abbreviation simply meaning 'the gas industry') is Russia's largest company. Its core interests are in gas and oil, but its assets range so wide – a television station, the former government newspaper *Izvestia*, a bank and even a football club – that it has been described as a state within a state. Indeed its former chairman Dimitry Medvedev gave substance to this description by exchanging his job for the presidency of the Russian Federation.

The Shtokman licence is held by Sevmorneftegaz (another abbreviation, for 'northern offshore oil and gas'), a wholly owned subsidiary of Gazprom. In 2008, the Russian parent formed an operating company with Total (France) and StatoilHydro (Norway). In return for a share of the profits – but not ownership of the gas – these foreign partners will provide much of the necessary technology, particularly for producing liquefied natural gas (LNG) that can be shipped out to North America in refrigerated tankers. However, if the proposed Nord Stream pipeline goes ahead, much of the gas will probably be pumped under the Baltic to Germany. It is supposed to start coming ashore in 2013–14, though given the enormous technical challenges involved, hundreds of miles offshore in a 1,000 feet of water, that looks optimistic. Russian weather forecasters have added a warning that while a warmer climate may generally make things easier, floating gas platforms in a stormy Barents Sea will paradoxically be in greater danger from drifting pack ice and million-ton icebergs running amok.

Other nations drilling offshore face similar physical problems as the arctic cold loosens its grip. And of course everyone will at the same time benefit, especially from the freer movement of shipping, whether they go in search of oil or one of the many valuable minerals also found in these high latitudes – nickel, copper, tungsten, tin, zinc, uranium, diamonds, platinum and gold. Some of these, for example Siberian nickel and Alaskan zinc, are already the basis of major industrial enterprises.

Oil and gas are nevertheless special, both because the scale of operations is so vast and because in our industrialised world energy translates so easily into political power. For the USA, the prospect of pumping more Alaskan oil means less dependence on unstable Middle Eastern suppliers. As of August 2009, Russia is already the world's largest exporter of both oil and gas. These exports have not only fuelled an economic resurgence, they give the Kremlin a powerful lever in dealing with European neighbours who depend on them. Being well placed to exploit new arctic reserves like Shtokman simply increases that leverage.

Turning off the gas

Members of the European Union already rely heavily on imported energy, and over the next 20 years that dependence – particularly on imports from Russia – is expected to increase. Russia already supplies a quarter of the EU's gas, a similar proportion of its oil, and even some of its coal. By 2030, almost two thirds of its imported gas could be coming from Gazprom.

To begin with, West European governments were unconcerned about this deepening relationship. There were obvious commercial benefits on both sides. If there were anxious voices, they were more likely to be heard in Washington than in Brussels, as US administrations viewed the scene from a strategic NATO perspective.

Attitudes changed after 2000 as President Putin moved to reassert the Kremlin's control over business oligarchs like Mikhail Khodorkovsky – whose Yukos group was ruthlessly broken up – and renegotiate the terms on which Western oil companies operated in Russia. And concern turned to alarm at the eruption of the so-called 'gas wars' with Ukraine and Belarus. Although Gazprom insisted it was merely putting its trading relationship with the two former Soviet republics on a proper commercial basis, the effects were brutally direct.

In January 2009, with heating demand at its winter peak, the Russian company closed the valves supplying Ukraine – thereby threatening transit supplies to much of central and southern Europe – while it argued with its neighbour about allegedly unpaid bills. The valves stayed closed for two weeks. Europe shivered; it realised that the Russian bear's commercial embrace could suddenly become a stranglehold.

Brussels bureaucrats responded to all this with renewed exhortation to seek 'energy security' through co-ordinated action to build a reliable partnership with Moscow. Individual EU governments reacted variously, depending on the extent to which they rely on Russian supplies and are prepared to trust Moscow politically. But there is a common interest – shared for different reasons by Gazprom – in creating alternative supply routes wherever possible.

Gazprom seeks diversification to provide direct access to Western markets and improve its bargaining position with potentially awkward transit countries. The Nord Stream project, possibly bringing arctic offshore gas direct to Germany without passing through the Baltic states, Belarus or Poland, is at least partly driven by this policy – which leaves former members of the Soviet bloc pondering an embittered Hobson's choice.

A different ball game

A retreating polar ice cap will change such strategic game plays in two ways. Given that the Arctic's particular economic resources are either finite (fossil fuels and minerals), limited (fish), or difficult to transport (timber), access to additional supplies will generally strengthen the hand of those who can physically and legally take advantage of it. And of course the remote northern communities directly involved may benefit in the process.

At the same time, commercial options will increase, especially where sea transport is concerned. It will become easier to ship mining supplies in or timber out. Oil companies that abandoned earlier hopes of moving oil and gas from arctic fields by sea (the Humble oil company's failed Alaskan experiment in 1969 with the ice-strengthened oil tanker Manhattan was an example) will find new routes opening after all. Being able to ship refrigerated gas by tanker as well as pump it through a pipeline – as the Shtokman operators in the Barents Sea currently intend – also gives valuable marketing flexibility. Overall, the arctic cost penalty will decrease.

But before all this can go much further, someone has to sort out who owns what in the arctic basin. Not an easy matter. The fact that such rich energy resources have to be shared out among at least five nations, straddling the old Cold War divide and with complex legal entitlements that are far from resolved, means that political squabbles – if nothing worse – are bound to arise. 'The arctic is ours' declared the flamboyant Russian explorer Artur Chilingarov as he planted his national flag at the North Pole. Just carried away by patriotic excitement? Maybe.

THE LAW OF THE SEA

It used to be that young men seeking escape from the restraints of conventional life ashore would 'run away to sea' – only to find themselves on a watchkeeping treadmill, or feeling the lash of naval discipline. And so it is with ships. While they do in one sense have the freedom of the seas, they are also constrained to a remarkable degree by complex rules that have evolved over centuries from the 'ordinary practice of seamen'.

Maritime law is comprehensive, transnational and almost pedantically detailed. The 'rule of the road' for ships, for example, is not just the equivalent of driving on the right or left and giving way at roundabouts. There are rules governing every conceivable situation in which two vessels might meet, cross or overtake one another. And exactly the same internationally accepted 'collision regulations' apply at night, even though only the perspective of a few precisely positioned navigational lights – red, green, white and yellow – serves to indicate an approaching vessel's relative movement.

Much maritime law is nowadays in the form of UN conventions, negotiated since 1948 when a specialised UN agency – the International Maritime Organisation – was set up, and coming into force as soon as a requisite number of nations have signed and ratified them. These conventions, for example that dealing with the *Safety of Life at Sea (SOLAS)*, embody ancient rules evolved piecemeal from hard experience: Samuel Plimsoll's load line, introduced in the nineteenth century to prevent unscrupulous ship owners overloading their (no doubt well-insured) vessels; the 'no cure, no pay' principle governing salvage; the provision of sufficient lifeboats, as demanded after the loss of the *Titanic*; the maritime pecking order specifying which type of ship gives way to another, including the familiar rule that 'steam gives way to sail' – though no yachtsman who valued his life would any longer insist upon it.

Equally important – and especially pertinent in the context of more widespread arctic navigation – are conventions setting new standards to suit changing modern circumstances: phasing out single-hulled oil tankers to lessen the danger of disastrous oil spills such as the Alaskan grounding of the *Exxon Valdez*; modifying the 'no cure, no pay' principle to reward salvors who minimise pollution even if they cannot save the ship; the prohibition of casual waste dumping; higher training standards for specialist jobs such as-in this context-ice navigation.

An altruistic tradition

Seafarers are also bound by law to go to the assistance of another vessel in distress. In my own experience they would do this anyway, law or no law, just as volunteers readily man lifeboats – though I suppose Richard Dawkins might argue by analogy with his 'selfish gene' that in such situations altruism and self-interest are inextricably mixed. Either way, as more cruise liners head north, the need to provide a search and rescue capability in remote polar regions is a serious worry, particularly for the Canadian government, confronting the prospect of an ice-free North-West Passage.

The tourist industry is always looking for fresh destinations to give their customers a sense of adventure. Now it has some big new attractions in the form of 'calving' icebergs, threatened whales and starving polar bears. Since 1989 the Russians have even managed to earn some badly needed foreign currency by taking sightseers to the North Pole in nuclear icebreakers – not such a spectacular voyage, incidentally, if when they arrived there was no ice, just an empty sea.

Increased realisation that the oceans contain all kinds of valuable resources has recently prompted new kinds of UN regulation to control not only surface activity but also economic exploitation of the seabed. So the good news about the struggle for control of the Arctic Ocean's emerging wealth – notably the gas, oil and fish – is that there already exists a solid body of international law to restrain it. The bad news is that all five coastal nations are in dispute with one or more of their neighbours as to precisely how that law should be interpreted.

Going to the ball

Until geologists caught the scent of offshore oil, the main law governing its exploitation was something of a legal Cinderella. The 1982 *UN Convention on the Law of the Sea (UNCLOS)*, sometimes known simply as 'The Law of the Sea Treaty', was an immense co-operative achievement, the result of nine years of complex negotiation, but it fell foul of Cold War rivalries between the USA and the USSR. With the USA firmly established as the world's predominant naval power, Washington was reluctant to concede authority to a multinational organisation – a prejudice which lingers to this day – and deeply suspicious of Moscow's motives for supporting such an arrangement.

Ironically, it was the Americans who had first asserted control over their adjacent continental shelf so as to protect their fisheries, and the potential for offshore oil production – a basic feature of the subsequent UN law. Yet when the 1982 convention was first drafted, Washington refused to sign, objecting particularly to the idea of an international authority to control deep-sea mining.

US attitudes have changed a lot since then. If the UN regime was going to be effective, the Americans wanted part of it. After all, they had one of the longest coastlines in the world. Increasingly, they realised that the benefits of an agreed legal framework to establish rights and settle disputes were likely to outweigh any loss of national sovereignty.

So most of the *UNCLOS* provisions came to be accepted in Washington as part of 'customary international law', even though the USA was not a party to it. And the Pentagon in particular urged its government to endorse the convention. The US Navy might ultimately have had the firepower to force its way across any foreign sea, but why ask for trouble when an international treaty was offering everyone the 'right of innocent passage' through other nations' territorial waters and strategically important international straits – for example, the Strait of Hormuz through which tankers must pass along the Iranian coast to load oil in the Gulf? Where the oil industry was looking for legal security, the admirals wanted freedom of manoeuvre.

Besides which, the political context changed completely with the collapse of the communist regime in Moscow. In 1994, changes to the convention were negotiated to meet Washington's remaining objections. The USA became a belated signatory, and it was assumed that ratification by Congress would soon follow. Yet at the time of writing, this final endorsement – though confidently predicted, and apparently supported by the Obama administration – had still not been granted. Republican senators in particular blocked it, citing the old loss-of-sovereignty argument. And this in spite of a clear statement from former president George W. Bush – the arch exponent of unilateral US power – that the Senate's approval would 'serve the national security interests of the United States, including the maritime mobility of our armed forces'. In this instance, his officials, so often sceptical of the UN, were even prepared to concede that it might be useful in the so-called 'war on terror'.

The 1982 *UNCLOS* treaty on the law of the sea replaced four earlier UN treaties. As with the *SOLAS* convention on maritime safety, this complex body of law both embodied traditional concepts and introduced new rules reflecting modern reality – for example, the ability to drill for oil offshore.

Within cannon shot

It had long been accepted that notwithstanding the 'freedom of the seas' maritime nations like the British and the Dutch were historically so keen to preserve, a coastal state was entitled to control its adjacent 'territorial waters'. Their extent was originally limited to 3 miles, because that was the range of cannon shot. So the UN lawyers' first task was to codify a more realistic limit of 12 miles – which most states were already claiming – plus a further 12-mile-wide 'contiguous zone' within which laws against smuggling or illegal immigration could be enforced.

These distances were to be measured from agreed baselines, which also define the 'internal waters' inside them. Not a contentious issue where the coastline is more or less straight but of great importance where it is deeply indented or consists – as in the Canadian arctic – of an extensive archipelago. Ottawa regards large parts of the North-West Passage as its internal waters, if only to protect them from pollution, and has stubbornly maintained its position when challenged.

But other governments, most notably the USA, have refused to accept this. The argument has been nagging away for many years, hitherto of little more than symbolic importance since the waters concerned were usually frozen. That will certainly change if the North-West Passage opens to navigation. In that case the US Navy, or any other ocean-going navy for that matter, will want this new arctic route from the Atlantic to the Pacific declared an international strait through which its warships can move freely, without requesting Canadian permission. Everything therefore turns on the various principles laid out by the *UNCLOS*, and the definitions attached to them.

Rights of passage

Foreign ships have few automatic rights within another country's internal waters. However, they do possess the crucial 'right of innocent passage' through the 12 miles of territorial waters outside the baselines – provided that passage is not 'prejudicial to the peace, good order or security of the coastal state'. A passing vessel is not 'innocent' if it test-fires naval weaponry, launches military aircraft, causes pollution or even just stops to fish. Submarines are required to navigate on the surface and to show their flag – though you can be sure they do not always do so.

Provided the rules are obeyed, coastal states are legally bound not to hamper innocent passage, and for merchant vessels, that is good enough. Warships, whether they fly the US, Russian or British ensign, are looking for something more – a 'right of transit passage'. This exists in major connecting waterways like the Strait of Gibraltar – and arguably on the new arctic routes that may be opening up – where surface warships can legally pass through in an armed configuration and submarines can slip through unseen.

On the shelf

However, there are two other related concepts embodied in the *UNCLOS* which when applied in detail to the Arctic Ocean are proving even more widely contentious. Both give coastal states an economic monopoly in their adjacent waters (which is where most of the undiscovered reserves of oil and gas described earlier are likely to be found). The first deals with the natural phenomenon of the 'continental shelf', the gently sloping extension of the mainland before it drops away into the deep ocean. The other has an ugly name – the Exclusive Economic Zone (EEZ) – which refers to a 200-mile-wide strip of sea beyond the coastal baselines. Taken together, they have hugely profitable implications for states which can take advantage of their complex rules.

Until now, that has hardly seemed to be the case in the Arctic. While the ocean remained frozen, arguments about who owned whatever was beneath its surface were largely academic. Rival national claims were debated but rarely resolved. The legal fraternity is never inclined to hurry over anything, and there was no obvious reason to do so here. The basic principles of coastal jurisdiction, as opposed to their precise application, were, after all, well established. But signs that the ice may suddenly relinquish its grip have given the legal process a new urgency and a much wider importance. Teams of state-employed lawyers are spreading out the polar charts and consulting the arcane wording of the 1982 treaty. If this law is a Cinderella, it is definitely going to the ball.

The original purpose of a 200-mile EEZ was mainly to enable states to control and protect their coastal fisheries, but it does also give them exclusive rights to exploit any oil, gas or minerals buried beneath – an immensely important provision as soon as deep offshore drilling became a practical proposition. Similar rights apply on the continental shelf. This extends arbitrarily out to 200 miles offshore, contiguous with the EEZ, and to a maximum of 350 miles (or 100 miles beyond the 2,500 metre depth contour) if the shelf naturally, geologically, stretches that far – a question of complex technical definition. Plenty of scope, as you can imagine, for legal dispute.

Who owns the North Pole?

When these geographical rules are applied to a map of the Arctic, and on the assumption that each of the five coastal nations maximises its claims (Iceland has similar rights, vital to its fishing

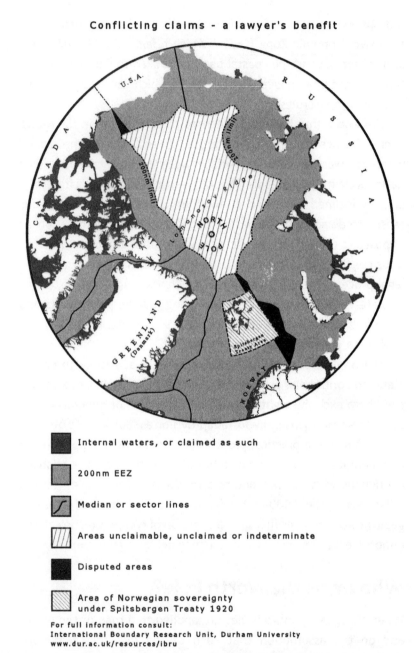

Map 4. Legal control of the Arctic Ocean's resources will be determined by the UN Law of the Sea.

industry, but outside the main oceanic basin), the result is dramatic. Almost the whole ocean is claimed by one nation or another. Only a few patches of relatively deep water on either side of the central Lomonosov Ridge remain definitely beyond any single state's jurisdiction – to be managed in legal theory by the International Seabed Authority, enjoying rather different weather in Kingston, Jamaica. Russia and Denmark even seem to think they own the North Pole!

What is more, the claims overlap in a number of important areas, usually because states cannot agree on how a dividing boundary should be drawn or because a pre-existing agreement – the 1920 Spitsbergen Treaty is a prime example – is believed to override the UN law. Where the continental shelf or EEZ has to be shared out between arctic neighbours, they have generally chosen one of two conflicting principles – a 'median line' reflecting the shape of the adjacent coastline, or a 'sector line' drawn straight from the North Pole to the extremity of their territory.

As you might expect, each government has tended to choose the method that suits its interests and then present this as self-evidently logical. In Stalin's day, for example, the Kremlin found no difficulty in delineating control of the Arctic Ocean – you simply applied the sector principle. Anything north of its coastline belonged to the Soviet Union. Two straight lines radiating from the North Pole – to Murmansk in the West and the Bering Strait in the East – established that half the ocean belonged to the USSR.

The same principle had been adopted by the old Imperial Russian regime and was vigorously reasserted by the Communist administration in 1926. It was used as a legal pretext to turn Norwegian sealers out of Franz Josef Land, and North American Eskimos off Wrangel Island, just inside the Bering Strait.

Since then, Norway, with the delicate firmness that characterises its dealings with its giant neighbour, has persuaded Russia take a more flexible approach. But the Barents Sea they share is still covered with a cat's cradle of disputed boundaries. Where fishing is concerned – vital to both countries' northern communities – they have for the time being just agreed to differ. A so-called 'grey zone' was established in 1978 within which each government has jurisdiction over its own vessels.

The sector-or-median line argument crops up again on the opposite side of the ocean, in the coastal waters between Alaska and the Canadian Yukon. Canada's claim is remarkable for citing a treaty signed so long ago, in the early nineteenth century, that the British still ruled the Yukon and Alaska belonged to Russia. Ottawa says it was agreed back then that the dividing line between the two territories should follow a meridian of longitude 'as far as the frozen ocean', which could in principle mean as far as the Pole.

After checking what this would mean in terms of EEZs, the American lawyers took a different view. To no one's surprise, they judged that a median line was more appropriate – it would give the USA control over an extra 7,000 square miles of potentially oil-rich seabed.

In the Bering Sea, shared by Russia and the USA, they seemed to have resolved competing claims in 1990 by swapping 'special areas', but at the time of writing, this agreement had not been ratified by a suspicious Russian parliament, which believed its fishermen were being robbed.

Less easy to explain is the long-running spat between Denmark and Canada over the status of Hans Island, little more than a lump of rock in the strait dividing Greenland from Canada's Ellesmere Island. The two governments are co-operating on geological surveys to back their respective claims concerning the Lomonosov Ridge, yet seem unable to sort out the ownership of this tiny outpost. What with government ministers landing on the island to pose beside their respective national flags, and troops burying bottles of whisky or brandy to show their opposite numbers 'we were here', there is a touch almost of comic opera about the affair. Admittedly, British marines got up to similar antics many years ago to establish

Britain's claim to the island of Rockall, off the Scottish coast, but in that case the potential gains seemed more obvious.

A Russian fist

For full-blooded theatre, however, one must look to the Russians. In August 2007, the flamboyant scientist-turned-politician Artur Chilingarov led a private expedition to plant a Russian flag at the North Pole, 4,300 metres (14,000 ft) beneath the frozen surface of the ocean. He personally descended to this frightening depth in the *Mir-1* submersible, previously used to explore the wreck of the *Titanic*. *Mir-2* followed (the name means 'peace') carrying foreign observers, including the Swedish pharmaceuticals tycoon who helped fund the expedition.

The 68-year-old Russian scientist – formerly head of the Soviet hydrometeorological service – admitted that the view down there was rather dull: no 'creatures of the deep', just 'yellowish gravel' which was sampled in the hope that it might bolster his government's claim to the Pole. Before surfacing, *Mir-1*'s mechanical arm reached out to place on the seabed a stiff titanium version of the red, white and blue Russian flag. The ascent almost proved more eventful, because although a hole in the ice had been cleared for the dive by the big nuclear-powered icebreaker *Rossia*, the submersible took an anxious 40 minutes to find it.

In truth, the adventure added little or nothing to Russia's scientific case for polar jurisdiction, but the symbolism was emphatic, the political impact unquestionable. President Putin telephoned from Moscow to offer his congratulations. Graphic accounts were carried across the world's press and television. Comparisons were even made with the Americans planting their flag on the moon.

Others tried to play down the hype, dismissing the Russian exploit as no more than a stunt. Canada's indignant foreign minister Peter MacKay protested on television that this was 'not the fifteenth century'; you could not claim territory just by planting a flag (or for that matter by dropping three different national flags from an airship, as Amundsen, Ellsworth and Nobile did on their flight over the Pole in 1926). Indeed the Russian foreign minister Sergei Lavrov commented diplomatically that while it was hoped the expedition would provide additional scientific evidence to support his government's case, the issue would be resolved in strict compliance with international law.

But for most foreign commentators, the message from Moscow resounded in Chilingarov's belligerent declaration that 'the arctic is Russian'. When he was reminded later of his foreign minister's more cautious statements, the old Soviet hand replied that Lavrov spoke as a diplomat. He, Chilingarov, would do everything possible to strengthen a Russian 'fist' to seize the North Pole.

The legal process for claiming anything more than 200 miles of continental shelf is subject to a deadline of ten years from the date on which later signatories of the *UNCLOS* (including the five states bordering the Arctic Ocean) ratified the treaty. The Commission on the Limits of the Continental Shelf (CLCS) requires detailed geological data demonstrating that the seabed being claimed is a continuous underwater projection from the mainland.

Russia, for example, lodged an early claim in 2001, but was told to go away and collect more information – hence, at least symbolically, the Chilingarov expedition. Norway submitted its claim in 2006. The Canadians and Danes are hard at work, mapping and surveying, and have until 2013 and 2014, respectively, to make their submissions. For the USA, which has already used its Coast Guard icebreakers to conduct extensive seabed surveys off Alaska, the clock is not yet ticking.

Political pressure, not just geology, will play a part in determining the outcome. But governments are obviously aware that jurisdiction over the seabed brings little benefit unless its wealth can realistically be extracted. And oil companies are not going to invest the vast sums involved in long-term offshore development unless they believe their exploration licenses are secure and their tax regime certain. This in turn puts a high premium, as far as the arctic states are concerned, on reaching an unambiguous settlement under international law.

Significant geology

Where control over the seabed is concerned, there are two key issues: how Spitsbergen should be treated and the geology of the Lomonosov Ridge.

The ridge, named after an eighteenth-century Russian scientist and poet, stretches for more than a 1,000 miles across the central ocean, skirting the Pole on the Pacific side. It is, if you like, an underwater mountain range almost, but not quite linking Siberia to Greenland – or more precisely, the Novosibirskiye islands on the Russian side and the north tip of Greenland, adjacent to Canada's Ellesmere Island, on the other.

The argument pursued by Russian and Danish lawyers is first a scientific one. Is the ridge geologically an extension of the mainland on either side, even though there is a trough of deeper water at each end? And if so, does this legally justify a claim for jurisdiction under the UN rules – or indeed as a matter of common sense? As one academic commentator remarked, the fact that the Scottish highlands were once joined to the Appalachians, before the tectonic plates drifted apart, could hardly be used by Scotland to claim sovereignty over North America.

The dispute over the Spitsbergen archipelago (Norway's Svalbard, the 'cold coast') is far more complicated. Not only is the geology subject to differing interpretations; jurisdiction under the 1982 *UNCLOS* rules overlays a contentious international treaty signed in Paris back in 1920, when the political map of the Arctic was quite different. And that treaty was in turn the culmination of an historic rivalry – and in some degree a practical partnership – between Russians and Norwegians.

The seventeenth-century Spitsbergen whaling boom which created 'Blubbertown', with its smoky boiling houses, was short-lived. In the fjords, the giant sea mammals were massacred. But walrus and seals remained, and on land there was a plentiful supply of arctic fox, reindeer and polar bears, which were trapped or shot by anyone tough enough to survive in such desolate conditions. Russian hunters were used to spending long dark winters in wooden cabins, and they moved in. One of them, Ivan Starostin, is said to have spent 39 winters on the islands – when bear and fox pelts were at their luxuriant best – and was buried on the shores of Isfjorden.

Large numbers of walrus and seals were still being taken throughout the nineteenth century, but increasingly Norwegians, not Russians, manned the ships or risked cold and scurvy by wintering ashore. It was this local predominance, a growing international scientific interest and the start of commercial coal mining at the turn of the century – for a few years there was something of a 'coal rush', involving American, British and German entrepreneurs as well as Norwegians – which eventually brought the issue of sovereignty to a head.

The Imperial Russian government was reluctant to concede Norwegian control. There was sporadic talk of some kind of international regime, possibly involving Sweden as well as Russia. But in the chaotic aftermath of the Russian revolution, Norway took advantage of the fledgling Soviet regime's weakness to strike a deal with the great powers of the day. Oslo's sovereignty was accepted in return for free international access to Svalbard's economic assets – notably coal. The USSR was not even invited to sign the treaty until 1934, on the grounds that its government had not been formally recognised, and Moscow only grudgingly – and temporarily – accepted the situation a year later.

At the end of the Second World War – when Soviet troops were in northern Norway helping to drive out the Germans – Stalin put pressure on a nervous government in Oslo to renegotiate the 1920 treaty or at any rate agree to joint military control. But the Norwegians managed to stall long enough for the whole strategic situation to be transformed by the onset of the Cold War and the establishment of NATO (with Norway as a key member) in 1949.

Fishing for facts

It was at the height of the Cold War confrontation that I made my only visit to Spitsbergen. I was already in the far north of Norway, reporting for the *Guardian* from NATO's remote border with the Soviet Union – a quiet countryside where the dividing line between the two countries was unobtrusive but in those days strategically vital. There was added public interest at the time because a South Korean air liner which lost its way over the Pole had recently been forced down on a frozen lake by Soviet fighters, killing two of the passengers. So the story I filed concerned the Norwegian border commissioner's efforts to extract an explanation for this shocking incident from his Soviet counterpart.

They had met – on another frozen lake – for a day's diplomatic fishing, sitting round a hole in the ice, vodka bottle to hand. After the usual exchange of courtesies, each man, as always, raised the question that most bothered his respective boss. For the Norwegian it was of course the fate of the air liner, a subject that evidently caused some embarrassment. Then it was the Russian colonel's turn – he wanted to know what the Norwegian authorities proposed to do about 'provocative' Chinese tourists in the area. It seemed a supremely irrelevant complaint even by Cold War standards, but on the way home through Tromso, still within the Arctic Circle, I went in search of a beer – to find that the local nightspot was run by Chinese.

I also discovered that a small airstrip had recently been opened on Spitsbergen. After centuries of isolation, when the archipelago was completely cut off from the mainland for eight months of the year, it was now possible to fly there twice a week from Tromso with SAS.

When the idea of an air link was first mooted, the Russians, typically, protested on the grounds that it might compromise the islands' demilitarised status. Once Aeroflot arrived alongside the Scandinavian airline, however, the new facility simply became yet another excuse for the familiar Russian game of challenging Norwegian authority that had been under way since the Paris treaty came into force half a century earlier. On one occasion the flight from Murmansk landed with a cargo of double beds – for wives of the permanent airport staff the Russians claimed were necessary to operate their monthly service. This domestic demarche ended in compromise – the Aeroflot manager could be accompanied by his wife, with or without a double bed, but wives of the other six staff could only make short visits.

Ice bears that go bump in the night

I booked my own flight at the end of April, reassured by a Norwegian diplomat from Oslo that spring had already arrived in Spitsbergen, so the reasonably warm clothing I brought with me for my original trip would be sufficient. The vast anoraks and fur-lined boots all the other passengers were wearing made me wonder about this, but the truth did not become clear until we began our descent to Longyearbyen, when the pilot announced a ground temperature of minus 10 degrees C and I looked down on a frozen sea in every direction.

In these latitudes, 'spring' hardly happens. The sea ice suddenly starts to break up – sometime in May – and it is summer. For people living just about as near to the North Pole as you can get, the seasonal transformation is deeply, emotionally welcome. I had arrived a

3. The old coal-loading berth, Longyearbyen.

little too soon to witness it, but in other ways this turned out to be convenient. The airport engineer's wife found me wandering through a hangar, and directed me to sleep in the 'hotel' – a construction worker's wooden hut – but not to open the door during the night because polar bears, or 'ice bears' as she called them, sometimes came sniffing round. Her advice sounded somewhat overdramatic until later I heard the story of the Austrian camper who opened his tent flap to a bear, only to be dragged off to a nearby ice floe and devoured.

Next morning, after breakfast in the miners' canteen, and with snow still fortunately on the ground, I was lent some proper clothing and a snowmobile with which to explore the stark, monochrome landscape. Simply observing 'the nature', my hospitable guide explained, was one of the main ways in which Longyearbyen's residents passed their time.

Things have changed a lot since then. Tourism has arrived in the Arctic. I am told there are now a couple of real hotels in Longyearbyen, and lots of scientists – including some Chinese ones! Spitsbergen has become an important international centre (20 different nationalities were represented at the last count) for vital research into the causes of climate change – 'greenhouse' gases like carbon dioxide and the dreaded methane – or the study of oceanic circulation and polar phenomena such as the aurora borealis.

The Cold War is long over, and with it much of the military paranoia generated by the fact that the channel between the Norwegian mainland and the archipelago (technically NATO territory, albeit demilitarised) was the only ice-free route by which Soviet naval forces based on the Kola Peninsula could reach the open Atlantic. Those Russian coal miners, who once heavily outnumbered their Norwegian counterparts, are no longer so important to the Kremlin. Their labour may still be 'glorious', as a placard outside their living quarters proclaimed, but this rundown post-Soviet settlement now registers potential economic interest rather than military preparedness.

Moscow keeps up the diplomatic pressure (reinforced by hints about the possible need for naval intervention) because there is still a lot to play for in terms of access to the Barents Sea's resources. Russia's northern communities desperately need its fish. If substantial oil or gas is found within the archipelago, Russian companies want their share of that too – and have announced their intention to look for it. So the overlapping, even contradictory, jurisdiction offered by the Spitsbergen Treaty and the more recent UN regime is a continuing source of friction.

Rights of access

The 1920 agreement gave Norway sovereignty, and administrative responsibility for things like fishery protection and pollution control, on condition that all other signatories – including the USA, Japan, China, the UK and eventually the USSR – had equal rights to exploit the islands' natural resources. The treaty lists the activities that are open to everyone – maritime operations (such as fishing, whaling and sealing), and industrial, mining and commercial operations. But crucially, this freedom only exists within the archipelago and its 4-mile-wide territorial waters (because this was negotiated long before the inception of 12 mile territorial limits and 200 mile EEZs).

As far as Norway is concerned, therefore, Spitsbergen is *not* surrounded by its own continental shelf, stretching out for 200 miles or more, to which foreign oil companies have access because their governments signed the Paris treaty back in 1920. On the contrary, Oslo's maps show the islands, distant though they may be, simply as part of a continuous continental shelf stretching out from the North Cape of mainland Norway, and way beyond towards the Pole – and the UN has supposedly given Norway exclusive rights to it.

If so, this is not a position that Russia – or indeed the UK – is currently prepared to accept. Add to this the residual dispute between Oslo and Moscow as to whether the valuable Barents Sea fishing grounds should be shared out by drawing a median line or by applying the Russian sector principle, and it looks as if the lawyers are going to earn their fees.

COLD WARFARE

The coinage of that useful term 'Cold War' has been credited both to the American financier Bernard Baruch and to George Orwell, the author of *Nineteen Eighty-four* – a more interesting attribution since the British writer used it with remarkable prescience to explore the implications of the atomic bomb, and the prospective stalemate of nuclear deterrence. For both men, it was the absence of direct hostilities between rival superpowers which made the war 'cold', not the climate. But as it turned out, the frozen arctic was geographically pivotal in that long confrontation.

Just as our atlases' familiar projections distort comparative sea routes between Europe and the Far East, they may also disguise the relative alignment of the continents. One glance at an azimuthal projection of the Arctic Ocean – a bird's eye view looking down from above the North Pole – shows that the USA and the former USSR are on opposite shores, with Canada serving as an unfortunate buffer between them. In the event of a nuclear conflict in the early years of the Cold War, many of the bombers attacking the other side's military bases would not have traversed the world in the way conventional maps suggest, rather they would have gone straight over the top. That is why in the 1950s the USA sought Canada and Denmark's co-operation in constructing the so-called DEW Line, a vast array of more than 60 early warning radars across the Canadian arctic from Alaska to Greenland (upgraded in the 1980s as part of the North Warning System).

The US Air Force had already established itself at Thule, in northern Greenland, where in 1951–3 thousands of workers were shipped in to build its northernmost airbase – precisely halfway between New York and Moscow. In 1968, this was the scene of a serious nuclear accident when a loaded B-52 bomber crashed on the ice nearby.

'The most valuable piece of real estate in NATO'

The Arctic was also strategically vital in Cold War naval operations, since many of the Soviet Navy's most powerful warships – submarines, missile cruisers and later aircraft carriers – were based on the Kola Peninsula near Murmansk to take advantage of the narrow tongue of warm ice-free water reaching round the North Cape of Norway. Their route from there to the open Atlantic lay through a web of NATO-controlled bases and listening posts, not least the former US Keflavik air base in Iceland (once described to me by a US admiral as 'the most valuable piece of real estate in NATO'). A Soviet convoy delivering missiles to Cuba plotted that southerly course in 1962, provoking a crisis which brought the two Cold War adversaries the closest they ever came to actual nuclear conflict.

Throughout the Second World War, allied convoys escorted by British warships had run the gauntlet of these same waters, under attack from German submarines, surface ships and aircraft based in Norway, to bring 'lend-lease' supplies to the Russians in Murmansk and Arkhangelsk. Terrible losses were sustained on both sides.

Hitler's forces even extended their military occupation to Spitsbergen, maintaining several radio stations on the islands to provide valuable weather forecasts. German U-boats made occasional forays into the Kara Sea – at the cost of several bent periscopes. In the summer of 1942, they were briefly joined by the heavy cruiser *Admiral Scheer*, which laid about her to some effect. And oddly enough it was a German warship, not a Russian one, which first made strategic use of the Northern Sea Route during the war, to move secretly right through from the Atlantic to the Pacific.

A helping hand

The German ship involved was the armed raider *Komet* – an innocent-looking merchant vessel disguising the fact that she mounted 6 inch guns and torpedo tubes, with a selection of national ensigns handy on the bridge. Superficially an extremely handsome cargo liner, she set out from Bergen in July 1940 during the brief period of the war when Stalin and Hitler were allies. Stored in her holds was enough fuel to go twice round the world.

At the entrance to the Kara Sea, she met the first of four powerful Soviet icebreakers hired to help negotiate the ice-bound straits that made this North-East Passage so impenetrable. With their assistance, and a couple of Russian pilots on board, she made excellent progress (according to Terence Armstrong, who provided this account, the Soviet government charged £80,000 for these services). The arctic fog was less persuasive than usual; the ice bottlenecks surprisingly clear. Three weeks and she was through – a record-breaking transit – and free to spend the next year roaming the Pacific.

From the German Navy's point of view, this unique redeployment was a great success. By the time the *Komet* returned to Hamburg she had sunk ten allied ships.

The Soviet Navy had no immediate reason to follow the German example, although a handful of its warships did later make the passage in both directions. The Russians were not actively fighting in the Far East until the end of the war and they had a railway to move troops. But for the few Russian sailors who knew about the *Komet*'s secretive voyage – just two weeks actually under way to complete the passage – it must have been deeply impressive. Many were old enough to remember another conflict, less than 40 years earlier, when the Imperial Russian Navy went to enormous lengths to move a whole fleet of ships from the Baltic to the Pacific without the option of a short cut across the Arctic.

Island of the donkey's ears

At war with Japan in 1904, Czar Nicholas II had decided to send a 'Second Pacific Squadron' across the world to relieve his First Pacific Squadron, trapped in Port Arthur and Vladivostok by Admiral Togo's forces. On arrival, the relief squadron's orders were to 'wipe the infidel from the face of the earth'.

A clear-enough instruction, but before Admiral Rozhestvensky could even begin to comply, he had to complete four battleships to lead his fleet, assemble the collection of obsolescent, inadequate craft that were to escort them, and then steam 18,000 miles to meet their enemy at the 'island of the donkey's ears'. The outcome was comic as well as tragic.

The admiral calculated that his 42 vessels would burn about 3,000 tons of coal a day. But unlike the British navy, they had no chain of coaling stations at which to refuel on their way round Africa and across the China Sea. Instead, he made an arrangement with a German firm to send colliers to rendezvous with the Russian warships at sea. It sounds like a modern naval replenishment exercise, but the reality was so unreliable that when he did manage to load coal, at Dakar, it had to be heaped all over his ships – even in the officers' baths. The British might have helped, but for a shocking misunderstanding which came to be known as the Dogger Bank incident.

Before the relief squadron sailed, wild rumours seem to have reached St. Petersburg that the Japanese might confront it before the ships even cleared the Baltic or the North Sea. For whatever reason, the Russians were in a thoroughly jittery state as they steamed across the Dogger Bank on the night of October 21, 1904, and suddenly picked out the dark silhouettes of the Hull trawler fleet. Convinced they had encountered a hostile flotilla of torpedo boats, they ignored trawlermen frantically waving fish in a searchlight beam to identify themselves and opened up with their 6 inch guns. One trawler was sunk and two fishermen died. The expedition might have ended right there, but after a half-hearted apology from the Czar, it was allowed to continue.

By the time the Russian warships reached Singapore, they were trailing a heavy growth of weed, no doubt reducing their ponderous eight-knot speed still further. There, the Russian consul hurried out to assure them that reinforcements – a handful of rusty old vessels grandly designated the 'Third Pacific Squadron' – would join them off the coast of what is nowadays Korea. It made little difference. In the Tsushima straits, a patrolling Japanese cruiser sighted the Imperial flagship *Suvorov* emerging from the mist on the morning of May 27, 1905, and by the next morning the battle was over. Only four Russian ships reached Vladivostok. Ten thousand sailors drowned.

Back in St. Petersburg, even the inept Romanov administration could hardly fail to learn some sort of strategic lesson from this disaster. Two new icebreakers were built to take another look at the arctic route, this time from the Pacific end. They set out in 1913 under Commander Vilkitsky and, after being defeated by the ice that season, were eventually successful two years later. But in truth Vilkitsky's hydrographic surveys merely confirmed how difficult it was going to be to turn the North-East Passage into a usable waterway.

The sailor Czar

If Britain, the USA and Japan, with their long coastlines, are natural sea powers, then Russia is the opposite. On a paper map, the Siberian coast is prominent enough, but until the development of modern ice-strengthened ships, and immensely powerful icebreakers, most of the sea might just as well have been a frozen extension of Russia's vast land mass. Murmansk was always the striking exception; Vladivostok another important outlet. Elsewhere, Russian sailors – Peter the Great most famously among them – faced a centuries-long struggle to reach warm water.

The Czar learned to sail in an English boat on Lake Izmailovo. As a young man, he visited Dutch shipyards and in 1698 London's naval dockyards at Deptford – where his rowdy companions virtually wrecked the house they were loaned – to learn for himself something of the latest shipbuilding techniques. Five years later his troops stormed the Swedish fortress at the mouth of the River Neva, and he began to build his new capital, spacious and beautiful, out there on the marshes. He also built a substantial navy. By the end of his life, it comprised 800 vessels and the Swedes had been driven from the eastern Baltic.

Peter also had his eyes on the Black Sea, controlled by the Ottoman Turks, and managed for a time to occupy a corner of its northern shores. Catherine the Great, who shared his appetite for imperial adventure, consolidated this breakthrough in 1770 with a spectacular naval victory over the Turks, eventually forcing them to grant her ships free navigation of the Black Sea and a right of passage through the Bosphorus and Dardanelles. A Russian squadron could sail the eastern Mediterranean – truly warm water at last. The Czarina's advisers even began to prepare plans for the partition of Turkey, but such ambitions planted seeds of suspicion in English minds which half a century later were to grow into a flourishing 'jingoism'.

In the event, the great maritime initiatives of Peter and Catherine were not followed through. Throughout the nineteenth century, Russian naval technicians did show a characteristic inventiveness, leading developments in mine warfare, torpedoes and high explosive shells, but any strategic ambitions were severely limited by lack of access to the sea, and further handicapped by a combination of official indifference and incompetence.

After the 1917 revolution, in spite of the cruiser *Aurora*'s symbolic bombardment of the Winter Palace in St. Petersburg, the Communist leadership considered its navy self-evidently subordinate to the heroic Red Army. During the Second World War, it was confined to coastal defence and operations in support of the land forces – the army's 'faithful helper' – although things might have been different had Stalin's pre-war programme to build a long-range navy been launched somewhat earlier. His government also initiated a programme of ship canal construction, beginning in 1932 with a link between the Baltic and the White seas, which eventually extended from the Arctic Ocean to the Caspian. That provided some internal freedom of movement for small freighters (up to about 2,000 tons) and naval vessels. But only in the second half of the twentieth century did Russia finally look to the open sea with serious intent.

Krushchev's herrings

The commander-in-chief who reshaped the Soviet Navy to become a major instrument of Communist imperialism – not merely scuttling between Leningrad and Murmansk or patrolling off Odessa, but confronting US nuclear forces in the Atlantic, showing the flag in the Mediterranean, the Indian Ocean and the Caribbean – was the remarkable Admiral Sergei Gorshkov. A reliable party man as well as a professional sailor, he was evidently a real Kremlin operator.

The admiral's first problem was that Soviet leaders tended to know better than the professionals what a modern navy needed. Stalin, a late convert to the usefulness of naval power, developed 'a curious passion for heavy cruisers'. However Krushchev, who appointed Gorshkov in 1955, thought cruisers were a complete waste of money; instead he wanted hundreds of missile submarines (in 1959 he boasted that his navy already had so many, 'we have enough to assign some of them to catch herring in the North Sea').

Gorshkov had to salvage some of the cruisers, restrain the submarine programme and gradually persuade his conservative, army-dominated defence establishment to fund the construction of a balanced, ocean-going force. Eventually, after some difficult doctrinal contortion, it would even include aircraft carriers. His domestic political campaign was helped by the increasing contribution nuclear-powered ballistic missile submarines were making to the USSR's strategic deterrence and the proliferation of other 'state interests' that might supposedly require naval support. Soviet merchant ships, for example, were aggressively infiltrating Western cargo-liner cartels, and Russian factory trawlers seemed to be everywhere, hoovering up other people's fish.

Gorshkov's navy suffered from a persistent sense of claustrophobic encirclement. Two of its fleets were still usually bottled up in narrow seas controlled at their entrance by members of NATO. Even the Northern Fleet's vessels could be shadowed with relative ease as they passed down the Norwegian Sea past Iceland. Stalin's post-war efforts to bully Turkey into revising the 1936 Montreux Convention which limits naval traffic through the Bosphorus (the reason, incidentally, that Soviet aircraft carriers were designated as 'cruisers'), and turn the Baltic into a 'neutral sea', had come to nothing; in fact less than nothing, since Moscow's pressure on Istanbul prompted the USA to bolster Turkey as a strong eastern pillar of the Western Alliance. However, once clear of these geographical bottlenecks, Russian warships developed a new freedom of manoeuvre – they began to emulate, for example, the Western naval practice of refuelling at sea.

By 1970, the hundredth anniversary of Lenin's birth, the Soviet commander-in-chief was able to lay on a worldwide naval exercise, pointedly code-named 'Okean', involving 200 ships supported by naval bombers and reconnaissance aircraft, ranging across the North Atlantic, the Mediterranean, the Indian Ocean and the Pacific.

'Sooner or later', he warned, 'the United States will have to understand that it no longer has mastery of the seas'.

This proud boast, like so many others, was silenced in the subsequent collapse of the USSR. Denied funds even to pay its sailors, large elements of the Russian navy were disbanded or left to rust at the quayside. By the turn of the century, no more than a remnant remained fully operational. Only in the past few years has something of the old expansionist spirit returned, and one of the factors prompting this revival is the prospect of the Arctic becoming a crucial new arena for competitive international activity.

A hint of violence

In May 2009, the Kremlin published a new security strategy listing the Arctic, along with the Middle East, Central Asia and the Caspian Sea as an area of potential conflict. 'In a competition for resources', the document warned, 'it cannot be ruled out that military force could be used to resolve emerging problems', adding that 'the existing balance of forces near the borders of the Russian Federation and its allies can be violated'.

The revised strategy replaced one published in 2000, shortly after Vladimir Putin became president. It was prepared by the secretary of the Russian Security Council, Nikolai Patrushev, former head of the FSB security service and who once planted his national flag at the North Pole (though arriving in his case by air rather than submersible). Not in general a belligerent document, the strategy's focus on Russia's arctic borders, in company with more familiar complaints about US military encroachment, attracted a good deal of attention. It codified a number of recent hints and signs that Russia was flexing its muscles to protect vital resources described by Putin's protégé and successor, Dimitry Medvedev, as 'our national heritage'.

It was Putin who in 2008 detected an unwelcome 'smell of oil and gas' about the West's covetous dealings with Russia, and in the following year, Medvedev's representative at NATO warned the alliance not to meddle in the Arctic, where he claimed it had no role to play. The Russian Navy had in the meantime hoisted its own signal by despatching a couple of major warships from Severomorsk in the summer of 2008 to patrol arctic waters for the first time since the collapse of the Soviet Union. Nothing aggressive about it, said a naval spokesman – merely 'in the interests of security'.

Several more specific reasons for such patrols were suggested at the time: to protect Russian fishermen in the Barents Sea, or perhaps the Russian geological survey teams heading for the waters west of Spitsbergen, where they might or might not have rights under the 1920 Paris Treaty. From their navy's point of view, it makes little difference. The important thing is to have some mission – any mission – to justify the expensive maintenance and deployment of those warships to the Moscow bureaucrats allocating military funds. Here, in other words, is one of those vaguely defined 'state interests' that have to be resurrected and identified if Russia's naval resurgence is to be sustained.

Similar arguments are of course being used by military planners bidding for resources in other arctic countries. The Canadian government has already been persuaded of the need to develop a new military base at Nanisivik in anticipation of the North-West Passage opening, and to build half a dozen ice-strengthened patrol ships; the US administration is under pressure to replace the Coast Guard's icebreakers. But Russian admirals are also playing a much bigger game – to recreate a blue-water navy that can show the flag on the wider oceans Gorshkov once claimed to have mastered.

Wakening bear

On December 11, 2007, Norwegian oil rig workers off Bergen were startled by the unannounced appearance of a Russian naval task force. It was led by a large vessel with an upturned 'ski-jump' bow, busy launching helicopters and fighter aircraft. For fear of an accident, the Norwegian oil men temporarily grounded their own helicopters. Any naval buffs among them would have recognised the distinctive silhouette of the *Admiral Kuznetsov*, the Russian Navy's only remaining aircraft carrier, out and about for the first time in years.

The 65,000 ton carrier was launched from a Black Sea shipyard in 1985 (designated as a 'heavy cruiser' to avoid falling foul of the Montreux Convention's rules) and still far from operational when the fall of the Berlin Wall signalled the collapse of the Soviet military bloc. She suffered the indignity of several political name changes – from *Riga* to *Leonid Brezhnev*, then *Tbilisi*, capital of Georgia – before appearing as the *Admiral Kuznetsov*, assigned to the Northern Fleet. There she spent most of the next decade idle at the dockside, or undergoing repairs, making only brief forays from her home port – for example, during operations surrounding the loss of the submarine *Kursk* off Murmansk in 2000. So her appearance on what turned out to be a 15,000-mile training cruise, taking in the Bay of Biscay and the Mediterranean, excited a lot of interest among naval analysts and journalists. Was the Russian bear waking from its long hibernation?

Russians themselves fostered this interpretation. A defence minister described the voyage as 'proof of Russia's serious intention to return to the world's oceans as a leading naval power'. Over the next year or so the deployment was indeed followed by other bluewater training exercises and courtesy visits; and those involving Syria and Venezuela (where the carrier's leading role was taken by the nuclear-powered cruiser *Peter the Great*) were clearly in support of Moscow's wider political initiatives. In February 2009 it was Irish fishermen's turn to be surprised by the *Admiral Kuznetsov*'s appearance off their coast – or rather alarmed, since she spilled a lot of poisonous oil while refuelling on her way back from another long deployment.

The Russian Navy still has a long haul to restore its former operational capability (Western naval observers pointed scornfully to the tug which usually followed the *Admiral Kuznetsov* around – just in case). But for the moment this is about image building as well as training, showing the flag, the domestic competition for scarce funds; and in this respect, the prospect of dramatic change in the arctic climate comes at just the right time. For Russian sailors especially, it will broaden the base of maritime activity against which a navy operates, whether that involves fishing, freight transport, tourism, drilling for oil or exercising the right of innocent passage. It may one day give them the long coastline they have always sought.

The silent service

In all this, submarines are an exception. They rarely 'show the flag'. Indeed most of the time they take enormous care *not* to show it. And if they are nuclear-powered – as are many operated by the USA and Russia, plus a small number of British, French, Chinese and Indian boats – the ice which stops powerful surface vessels in their tracks is not necessarily a problem. For them the North-East and North-West passages, or at any rate their deep water equivalents, have always been open.

Whereas the conventional diesel-powered submarine must surface periodically to charge the batteries that drive it when submerged, a nuclear boat is not dependent on a periodic supply of air. It can stay submerged indefinitely, with enormous power at its disposal – the sort of power that can, and has, dragged a trawler down by its own nets. So whereas for one type of submarine an extensive ice pack is a potentially deadly trap to be avoided, for the other it may be a convenient hiding place. Combine this with the fact that during the Cold War, many of the former Soviet Union's most formidable submarines were based at Severomorsk on the Barents Sea, eventually able to target their ballistic missiles on

US cities without ever leaving home waters, and it is clear that in these strategic terms, the Arctic has long been an active arena.

Submarine warfare is an essentially secretive activity – hence references to the 'silent service'. Unable to see unless they raise a tell-tale periscope to scan the horizon, submarines rely heavily on sound to feel their way around – and as it happens, sound travels extremely well under water. Once submerged, everything is startlingly audible, from the rhythmic thud of a ship's propeller to the clank of an anchor chain, from a whale's melancholy song to the crackle of a shoal of snapping shrimps. Different types of surface vessel – even individual ships – can be identified by their sound 'signature' as they pass by above. The same goes for other submarines, unless they proceed very slowly – and therefore very quietly – or manage to hide among water layers of varying density which deflect sound.

The submarine's military task is to sink an enemy's surface ships and, if it can find them, hostile submarines which may threaten its own shipping. Practising for this stealthy form of warfare, even in peacetime, involves exploring the approaches to other countries' ports and naval bases, lying silently in wait for other vessels, or trailing them without giving away one's own presence.

It follows that throughout the Cold War there was constant invisible activity beneath the waters of the North Atlantic and the Barents Sea involving NATO and Soviet submarines, often in close proximity. Indeed there were a number of collisions, rarely acknowledged by either side unless disclosure was unavoidable for some political or technical reason. And to a lesser extent, the same underwater games are still being played out. When the big Russian cruise missile submarine *Kursk* sank during naval exercises in the Barents Sea in 2000, it was reported that two US boats were in the area listening to what was going on, a fact which generated a flurry of claims and counter claims as to whether the Americans had been involved in the accident.

Polar snapshots

Back in the 1950s, therefore, the development of nuclear propulsion for submarines prompted a small revolution in naval operations across the Arctic. The Americans' first nuclear boat, the USS *Nautilus*, put to sea in 1955, and within a couple of years, she was venturing under the polar pack ice. Her first attempt to complete a full transit, from East to West, in June 1958, was blocked by deep ice in the Bering Strait – there was not enough clear water between the bottom of the compacted ice and the sea bed. But a few weeks later she made another attempt, using a different channel, and on August 3, passed directly over the North Pole on course for the Atlantic. Less than a year later the USS *Skate* actually surfaced at the Pole, to take the first of those now familiar photographs of a bulbous black hull silhouetted against the intense whiteness, with a few clumsily clad sailors wandering around on the ice, astonished at their situation.

4. HMS Tireless at the North Pole – a routine operation!

A similar picture of a Soviet submarine, the *Leninsky Komsomol*, at the Pole in 1963, carried a reproachful caption claiming that Russian nuclear boats 'were in the high latitudes of the arctic basin considerably earlier than the Americans'. If so, which sounds most unlikely, the Russians had only their habitual secrecy to thank for not receiving credit where it was due.

Nowadays such polar exploits are more or less routine. In 2004, a British submarine HMS *Tireless* surfaced at the North Pole in company with the American boat USS *Hampton* as part of a series of joint 'ICEX' training exercises. Three years later *Tireless* was once again deep in the Arctic Ocean, accompanied by the USS *Alexandria*, when an exploding oxygen generator killed two of her young sailors and forced her to surface – a reminder that even supposedly routine trips are not free from the dangers inherent in submarine operations.

In the course of her winter transit, *Tireless* had been using her sonar to measure the thickness of the pack ice above. 'Sonar' is short for 'sound navigation and ranging'. It is the modern naval successor to the ASDIC whose rhythmic echoing sound became familiar from war movies like *The Cruel Sea*. Merchant vessels and yachts also use similar equipment, known simply as an 'echo sounder', to measure the depth of water under their keel. All three systems work by sending out a directional sound pulse whose echo is timed – bouncing back from the sea bed or another vessel – so as to calculate depth or distance.

A naval submarine relies heavily on sound, to both navigate and fight. And in the Arctic, sonar is doubly useful because by directing it upwards, submariners can check the contours of the pack ice above them just as an echo sounder shows the varying depth beneath the hull. If they intend to surface through the pack, this extra information is vital. Underwater ridges of compacted ice must be avoided, and although in extremis suitably modified submarines can break through more than a metre of smooth young ice, they prefer to find a patch of open water.

Coming up for air

Peter Wadhams, Cambridge professor of Ocean Physics and one of the world's leading experts on the changing arctic climate, was on board the smoke-filled HMS *Tireless* in March 2007 when she crashed up through the Alaskan ice. He had joined the trip to take measurements of ice thickness on the Canadian side of the Ocean, using the latest multibeam sonar – because it is the dwindling thickness of the cover, not just its extent, which determines how soon the summer ice is likely to disappear.

There are other, less adventurous ways of taking such measurements: electromagnetic devices mounted on helicopters or aircraft; satellites equipped with lasers or radar. But sonar is the most direct method, and has already been used over several decades to provide comparative data. Freed from the censorship of the Bush administration, the Americans are co-operating in the research, and have recently released ice transit records and satellite surveys to fill gaps in the picture of climate change. To this small extent, swords are being turned into ploughshares.

East by North

It is a commonplace but nonetheless remarkable fact that standing on the seashore, the curvature of the earth is actually visible – a ship only a few miles offshore will begin to be 'hull down'. Equally remarkable is the determination with which our seafaring ancestors, having grasped that their apparently flat earth was actually spherical, set out with only the clumsiest of sailing vessels, and no charts, to put that understanding to practical test.

If China and the Indies could be reached by sailing eastward from Europe round Africa, it must also be possible to do the same by going west-about. Columbus and Cabot might well have done so had the Americas not blocked their path. Magellan found a way round by turning south and following the coast for thousands of miles until he found the strait that bears his name. Drake, sponsored by England's Queen Elizabeth I, followed in his wake. But the Elizabethans were already turning their attention to the only other route by which they might be able to avoid that great obstacle – through the Arctic. By doing so, they would also be avoiding southern waters controlled by their powerful rivals, the Spanish and the Portuguese.

'As plausible as the English Channel'

The Elizabethan explorer Martin Frobisher, a buccaneering sea dog if ever there was one, knew that although the Russia Company's ships had already traded successfully to the White Sea, attempts to complete a North-East Passage had been blocked by ice at the entrance to the Kara Sea. Instead he favoured the possibility – indeed near certainty as far as he was concerned – of a North-West Passage. For him it was 'as plausible as the English Channel', and in 1576 he persuaded the company to fund an expedition to find it. The Queen herself waved from her window as his two small ships – the larger of them less than 50 feet long – set off down the Thames.

In the event the expedition found no more than the first of many dead ends that characterised a search lasting sporadically for another three centuries. However, Frobisher did return from his first voyage with black ore thought to contain gold, and was hurriedly sent back by his financial backers to find more, and then again for a third time. By now, finding a North-West Passage was secondary to making money by establishing a permanent settlement to exploit the 'gold', but a combination of gales, ice and some painful encounters with the local Eskimos put paid to such ambitions. The tons of black ore he laboriously mined and loaded turned out after all to be worthless pyrites.

Gin on the rocks

The gallant captain's geographical legacy was to have his name attached to a small inlet – Frobisher Bay – on the southern tip of Baffin Island. And the story of the subsequent search for a North-West Passage can be read almost chronologically from the names of the numerous other bays, straits and sounds named after his successors as they slowly probed deeper into the archipelago – Davis Strait, Hudson Bay, Baffin Bay and so on. Among the exceptions were the ancient identity of Greenland – a piece of public relations by the Vikings, who were there long before the Elizabethans – and the Boothia Peninsula, named at a much later stage after the London gin manufacturer who paid for a nineteenth-century expedition.

The Dutch had meanwhile followed English pioneers round the North Cape of Norway in search of their own trade with Russia and a North-East Passage to the Indies. It was the merchants of Amsterdam who funded the great Willem Barents, commanding one of three vessels heading north in 1594, and already prepared not merely to feel his way eastwards along the mainland shore, but to strike out round the farther end of Novaya Zemlya. Two years later, now sailing as pilot to Jacob van Heemskerck, he headed almost due North from the Norwegian coast, first sighting what they called 'Bear Island' (because they landed there and killed a polar bear) and then, when the persistent fog suddenly cleared, the jagged mountains on the north-western tip of 'Spitsbergen' – the descriptive name which soon replaced a simple charted reference to 'The New Land'. A sun sight put their latitude at 79 degrees, 49 minutes N.

Before returning to Amsterdam, Barents persuaded his captain to make another effort to enter the Kara Sea round the northern cape of Novaya Zemlya – a long easterly detour that ended disastrously, but also triumphantly. They had barely rounded the cape when ice blocked their path, and by mid-September they were hopelessly trapped at 'Ice Haven' on the eastern side. Not much of a haven. Their ship was squeezed and crushed by grinding pack ice. Fortunately, they found a lot of driftwood on the beaches – enough to build a large wooden hut with a central chimney under which they could maintain a fire to alleviate the deepening cold.

Throughout the long dark months – the sun did not reappear above the horizon until January 27 – the Dutchmen were constantly pestered by inquisitive foxes and highly dangerous bears. But somehow they survived to become the first Europeans successfully to spend a winter in the uninhabited Arctic.

We know their story – where many other stories were lost with men who died – because in the following June they were able to launch their small open boats and struggle south to the Russian coast, and from there back to Holland. Barents himself never made it. He died of scurvy on a drifting ice floe where they temporarily took refuge. Having discovered and named the distinctive features of the Barents Sea, his own name is rightly attached to it – even on Russian maps.

Through the early years of arctic exploration, European sailors dreaded being trapped through such a long winter, facing cold, scurvy or simple starvation. Although the whaling companies tried almost from the start to persuade men to remain ashore through

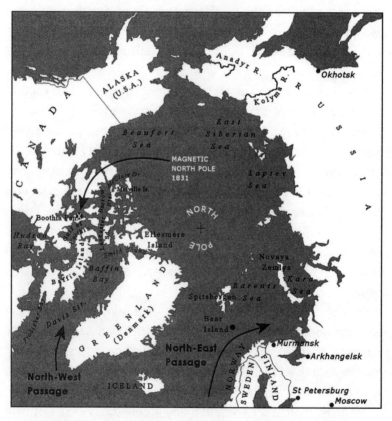

Map 5. Two Arctic Routes from the Atlantic to the Pacific.

the winter, 'cooking' blubber, it was only much later that expeditions went out equipped and fully intending to spend several seasons among the ice – and still they sometimes died.

The inevitable shipboard tension was at its worst when a ship reached the point of no return, and her ambitious captain had to decide whether it was worth pressing on, with or without the consent of his crew. Suspicion and superstition easily took hold among sailors living in cramped conditions, poorly fed and probably without the information to weigh the dangers for themselves. Mutiny threatened more than one voyage, and in the case of Henry Hudson, it saw him actually cast adrift like Captain Bligh of the Bounty.

Wishful thinking

English sailors searching for a North-West Passage would start by sailing westwards to pick up the southern tip of Greenland – Capes Farewell and Desolation, never an encouraging sight – and then work their way up into the bays and channels that lay beyond it. Reading their accounts when one already knows what is there can be quite illogically frustrating. Why did such and such a navigator spend months sailing down a blind alley when a clear channel obviously – in retrospect – lay just to the north or south? Perhaps it just *looked* promising if you were there on that day, as the land fell away, or maybe they were following a lead suggested by someone else. In the case of the notorious Zeno brothers' map, supposedly based on some ancestor's voyage, they were confused by blatantly false information.

In truth nearly every voyage – and there were dozens over the almost three centuries it took to solve this great navigational mystery – contributed something to the chart, if only negative confirmation that the passage was not to be found in that particular corner of the archipelago. Shortly after Frobisher's return, for example, John Davis used the strait that now bears his name to probe northwards to the west of Greenland. He took a long look at Cumberland Sound, a deep inlet that almost – but not quite – divides Baffin Island, and then hurried straight past the noticeably turbulent entrance to the vast inland sea Hudson would later discover.

Henry Hudson's contributions to the search were not immediately useful, but would eventually prove immensely profitable. On his first voyage for the Russia Company, in 1607, he was under orders to sail his tiny sloop directly towards the North Pole to test a remarkably persistent theory that at the very highest latitudes, the Arctic Ocean would miraculously be warm and ice-free (one fanciful sixteenth century map by no less a cartographer than Mercator shows the Pole as an enormous rock from which channels radiate). He turned back disappointed, at a latitude of more than 80 degrees N, but not before discovering that the shores of Barents' Spitsbergen were teeming with whales, seals and walruses. His report to that effect was the signal for the great whaling boom that followed.

Spared the gallows

Hudson's arctic voyages all seemed to be dogged by the threat of incipient mutiny, and it was through mutiny, the subject of a famously gloomy painting in London's Tate Gallery, that his life ended. On this last trip he was specifically intent on finding a westerly passage, prompted by the fierce 'overfall' Davis had noted at the entrance to what we now call Hudson Bay – suggesting a strong tidal current produced by a large body of water. The bay he found inside the strait was so large that he could only explore the eastern shore before the ice closed in, and having set out in April with just six months' provisions, his crew faced a long, desperately hard winter. Shortly after they broke free in the following June, a mutinous group took control, dumped four of their sickly shipmates in an open boat along with the captain, his son and a couple of loyal sailors, and headed for home.

Among those quickly sent back to exploit Hudson's discovery were a surviving member of his crew, Robert Bylot – presumably spared the gallows because of his valuable knowledge – and the navigator William Baffin. They could find no exit from the bay to the north-west, but that was an insignificant failure against the subsequent success of the Hudson Bay Company. Many of its employees later made important contributions to the exploration of the North-West Passage by land and river. More than that, they developed a profitable, wide-ranging trade in furs that eventually did much to form the emerging shape of British Canada.

William Baffin was a brilliant navigator who developed his skills on some of the early Spitsbergen whalers. In his time, he was working at the cutting edge not merely of geographical but of scientific exploration. He was, for example, the first to take a lunar sight at sea – rather than the more straightforward sun sight – to establish his position.

On his last arctic voyage in 1616, still with Bylot, Baffin found a way through the pack ice to the northernmost reaches of what became his Bay. At the far end they penetrated Smith Sound, at more than 78 degrees N, his ship's log recording an unprecedented westerly compass variation of 56 degrees – because the needle was pointing not at the true Pole but at the Magnetic Pole, whose exact location in the archipelago would not be discovered for another two centuries.

By that time, Baffin's achievement – like the much earlier Viking voyages to North America – was effectively forgotten. On his way home, he had passed Lancaster Sound and named it after his patron – not realising it was the entrance to the main North-West Passage. But the English would not resume their search in those waters until the early years of the nineteenth century.

In the absence of formal exploration by sea, land-based hunters and fur traders nevertheless gradually extended their outposts across the Arctic, both east and west, filling in blank spaces on the map as they encountered a succession of great northward-flowing rivers. Towards the end of the sixteenth century, the powerful Stroganov family funded the Russians' first major effort to explore and control Siberia – a Cossack military expedition led by Yermak Timofeyevich, the folk hero after whom the first big Russian icebreaker was later named. He died wounded in an early battle with the Tatars, but generations of Cossacks followed him, finally reaching Kamchatka and the Pacific in 1697. On the opposite side of the Pole, the Hudson Bay Company had been granted its monopoly by King Charles II in 1670. And while its methods were quite different, the wide network of trading posts it established eventually served similar imperial purposes.

The end of the Cossack campaigns roughly coincided with the young Czar Peter taking full control in Moscow. He listened to the returning mercenaries' tales of exploration and conquest, and noted – as a sailor – that they had still not penetrated to the far north-eastern corner of Siberia, home of the notoriously belligerent Chukchis, and could not therefore settle the vital question of whether Asia was connected to America. With so many other preoccupations it was many years before he acted on his curiosity, but shortly before he died in 1725, the Czar launched the 'Kamchatka Expedition' to find the answer.

Logistical nightmare

The man to lead the expedition was Vitus Bering, a Dane serving in the Russian navy; the task itself was a logistical nightmare. Siberia is so vast that even in more modern times Russian soldiers posted to the Far East would say their goodbyes on the assumption they would never return. Before Bering could begin his exploration, his men had to haul tons of supplies, materials and equipment through thousands of miles of forest, across rivers and swamps to Okhotsk, and then build his two ships.

Three years after leaving St. Petersburg, Bering was ready. But at this critical point, the drive and determination that enabled him to reach the Pacific seemed to desert him.

The challenge was to sail his small vessel the *St. Gabriel* from the Anadyr River on the south of the Siberian peninsula to the Kolyma River on the northern side, so as to establish that there was a strait separating Asia from America. In the event, he came across some local Chukchis who assured him that a sea passage did exist, though blocked by ice on the northern shore; so rather than risk being trapped there, he just took a quick look round the East Cape and turned back. After an equally cautious and unproductive venture to the eastward the next summer, he rushed – a relative term – back to St. Petersburg to report to the Czarina Anna, having been away for five years.

If the Dane thought he had completed the task set him by Peter the Great, it did not look that way in Moscow. He was soon on his way back to Okhotsk with orders to lead the 'Great Northern Expedition', a more ambitious programme of exploration not just to confirm the existence of an outlet from the Arctic Ocean to the Pacific, but to chart the whole coast of Siberia, section by section. And they almost succeeded.

It took eight years of exhausting, dangerous work, often in appalling conditions, using the great northern rivers to launch survey parties into an Arctic Ocean where their clumsy craft were hampered by persistent ice and fog, or they were forced to take to sledges. One foolishly enthusiastic young naval lieutenant took his bride along as part of their honeymoon, only for both of them to die as they tried to make it back to their winter base.

Bering's specific job remained to prove the existence of the strait that now carries his name and discover what lay on the far side. He did eventually reach America, but would allow the ship's German naturalist just one day ashore to scurry round recording everything he could find before setting sail again. Even then the onset of autumn proved too much for their — no doubt desperately unweatherly — ship, which ended up stranded on an island off Kamchatka. Bering died of scurvy before the winter was out.

It was left to that consummate seaman Captain James Cook, who learned his trade in the North Sea collier brigs, to confirm the basic geography about which his Danish predecessor had only speculated. On his third circumnavigation in 1778, hoping perhaps to return from the Pacific by way of a North-West or North-East passage, Cook followed the northern coast of Alaska until the ice pack blocked him at Icy Cape, and doubled back to trace the Asian shore as far as North Cape, where he met more ice, before retreating southward through the Bering Strait. There, the mist which cursed so many arctic voyages lifted, to reveal simultaneously both continental shores.

Filling in the blanks

In the early nineteenth century, freed from the military constraints of the Napoleonic wars, the British navy returned to the search for a North-West Passage to China. In 1817, William Scoresby, a Whitby whaling captain who combined hard experience of arctic navigation with a scientific interest in phenomena like the Magnetic Pole, alerted the Royal Society's Sir Joseph Banks that ice conditions to the west of Greenland were unusually easy — it might be a good time to take another look.

Scoresby knew how unpredictable and dangerous ice could be. He did not himself believe the North-West Passage — if it existed — would ever be a practical commercial proposition. But the politically astute Admiralty Secretary John Barrow sensed that this was indeed the moment to launch a campaign. England would be 'laughed at by all the world' if she did not capitalise on her earlier exploration, he declared. And if *she* did not, the Russians might.

In effect the British naval explorers were starting almost from scratch, scarcely believing what Baffin had discovered 200 years earlier, and still deluded by the idea that an easier, ice-free route might even be found straight across the Pole. Their chart of the Canadian archipelago was blank, except where Hudson Bay Company employees Hearne and Mackenzie had followed rivers northward (the Coppermine and Mackenzie rivers, respectively) to reach an icy shore. Beyond that, just empty 'Polar Sea' was shown – which we now know is thick with islands.

Two pairs of ships set out in 1818 with, among their crews, men whose names were to recur throughout this phase of arctic exploration – Ross, Parry, Sabine and Franklin. The first ships headed due north to end up battered by ice on the coast of Spitsbergen. The second pair, commanded by Captain John Ross, rediscovered Baffin's Bay and, most importantly, sailed 50 miles deep into Lancaster Sound on its western shore, until the captain 'distinctly saw the land' enclosing it ahead of them.

Sadly for Ross's reputation, his eyes had deceived him. No one else saw the mountainous barrier he later described, and especially not Lieutenant Edward Parry, in the other vessel.

It was the charismatic young Parry who was given the chance to return the following year, and this time he penetrated westward hundreds of miles beyond the imaginary mountains, along what is now known as the 'Parry Channel', until he was stopped by the impenetrable pack ice blocking the McClure Strait. In a single voyage he had almost completed the passage which, if the arctic melt goes far enough, may one day become the simplest, deep-water route to the north-west. But he was not to know that. Instead he was forced to winter in a bay on Melville Island – and did so with astonishing success, even comfort, given that his were the first British naval vessels to do so. He organised theatrical performances, playing the fiddle himself; he even published a newspaper, the *North Georgia Gazette and Winter Chronicle*. When the sun returned, he organised a sledging trip – another first for the navy – across the island. His colleague Edward Sabine, an expert in magnetic observations as well as the *Chronicle*'s editor, had also made an important scientific and navigational discovery – that their ship passed to the north of the Magnetic Pole – for a while, the compass needle pointed south!

On its return to England the next summer, the expedition was given a heroic welcome. By skilful seamanship, courage and management, Parry had made arctic exploration look almost easy – a spectacularly wrong impression. It was another 30 frustrating years before John Franklin's last expedition – in itself a complete disaster – led to the discovery of a North-West Passage. Parry made another two voyages, the first finding an outlet – albeit unusable, since it was frozen solid – from the far corner of Hudson Bay, and the second ending in the loss of one of his ships, the *Fury*, crushed by ice in spite of its massive construction.

'The man who ate his boots'

On land meanwhile, Franklin and his companions had suffered terrible hardship trekking west and north from the Hudson Bay Company outposts, with the help of Indian guides, to chart the coastline stretching either side of the Coppermine and Mackenzie rivers. From his desperate efforts to ward off starvation, he became popularly known as 'the man who ate his boots'. Years of painstaking, dangerous work eventually demonstrated that there was at least a continuous ice-strewn channel along this southern edge of the archipelago. Later he was to lose his life struggling to break through to that same channel from the north.

In 1829 John Ross, no doubt hoping to redeem his previous failure, persuaded the wealthy London gin distiller Sir Felix Booth to fund a voyage that would probe beyond the desolate beach on which Parry was forced to abandon the Fury. Ross added to the usual difficulties of such an expedition by choosing a paddle streamer which could only manage three knots under power (the cumbersome, leaking steam engine was later dumped ashore), but with it he did manage to penetrate deep into the bay he dutifully named the 'Gulf of Boothia' – far too deep as it happened, because it proved impossible to extract his vessel from the ice. After spending a record four winters in the arctic, the crew finally escaped from this trap in open boats left behind by Parry and were picked up, almost miraculously, by the whaler *Isabella*, out of Hull. Ross had once commanded her!

This voyage did restore the captain's reputation – Ross was knighted for it. He had drawn new lines on the chart, though even now he misleadingly marked two important small channels as 'bays' which led nowhere. Most of his crew had survived. He had demonstrated through friendly contact with the Eskimos that their knowledge of the local geography could be relied upon. And most importantly from the Admiralty's point of view, on one of the long sledge journeys made by his nephew James Clark Ross, who had joined his uncle for the adventure, the exact location of the North Magnetic Pole was discovered.

Dip circle

In his account of the discovery, the younger man apologised for the fact that the long-sought-for Pole did not turn out to be 'a mountain of iron, or a magnet as large as Mont Blanc'. It was on the flat western coast of the Boothia Peninsula that his 'dip circle' – a sort of vertically mounted compass to record inclination from the horizontal – finally dipped to 90 degrees. He built a stone cairn, still visible, to mark the spot. But even as he did so the magnetic centre was moving, albeit almost imperceptibly, to the north-east.

Not realising that the expedition was about to reappear, a search party led by George Back had set out in 1833 down the Indians' 'Great Fish River' (now bearing his own name), which was believed to reach the sea roughly where the missing ship was heading. In the event no rescue was required, but another blank section of the chart was filled in, and a few years later the remarkable Thomas Simpson, an employee of the Hudson Bay Company who could cover 50 miles a day on snow shoes, reached the same area from the westward by way of the Coppermine River.

It was in this confusing, ice-choked corner of the archipelago around 'King William Land' – in reality an island – that the search for the North-West Passage reached its dramatic climax. In 1845 the Admiralty decided to mount one last, superbly equipped expedition to solve this maddening navigational puzzle. And the central, ultimately tragic role was played by the arctic veteran Sir John Franklin, chosen to lead it in spite of his advanced age.

Creating a Victorian legend

The story of the last Franklin expedition is not merely an account of historical events. For the Victorians and their naval propagandists it became a legendary tale of stoicism in the course of duty, imbued with moral significance. As a prolonged drama, set in a savage arctic wilderness, it caught the public imagination. Franklin became a national hero – not in the same league as Nelson or Wellington, but comparable to, say, Gordon of Khartoum or Florence Nightingale. 'Lord' Franklin became the subject of at least one popular ballad. Pubs were named after him. His exploits were depicted in narrative paintings like Thomas Smith's *They Forged the Last Link with Their Lives* in the National Maritime Museum at Greenwich, showing exhausted men desperately trying to haul a heavy open boat across the ice.

And the devoted Lady Franklin gave the story an extra impulse. Where the Admiralty eventually gave up the search for the lost expedition, she would not, being doggedly determined to solve the mystery of how the 129 men had died, and what her husband in particular had achieved.

5. A Victorian legend: John Franklin.

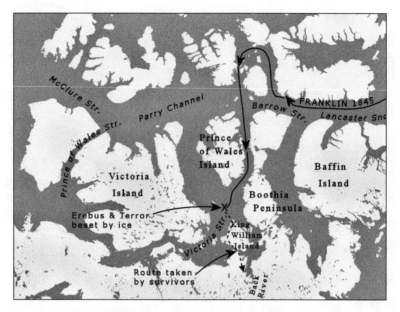

Map 6. Franklin's Last Voyage.

Jane Franklin was a strong minded, adventurous and pretty woman. She met her husband as a friend of his first wife, who died while he was away in the Arctic on one of his earlier land expeditions. His first major posting after this new marriage was as governor of the penal colony of Van Diemen's Land (now Tasmania), where she played an active part in developing the new community. And she did not sit about moping when he set out on his last voyage, instead dashing off on a quick tour of the West Indies and North America. On her return, the same energy was applied to organising a search for him as it became more and more clear that something had gone badly wrong.

Home from home

Yet the expedition which set sail from Woolwich in May 1845 could hardly have been better prepared. Franklin's two ships, the *Erebus* and *Terror*, were both what the navy called 'bombs' – powerfully built to withstand the stress of mounting heavy mortars, and further strengthened with iron sheathing, doubled planking and extra beams to withstand crushing. Auxiliary power was provided by steam engines (one bought from the local Greenwich railway) driving propellers, which must have made tight manoeuvres through the pack ice much easier.

The ships were provisioned for three years, including state-of-the-art canned meat – some of which was later found to be dangerously contaminated – and large quantities of lemon juice to prevent scurvy. (It was Captain Cook, back in the eighteenth century, who realised the effectiveness of citrus fruit in warding off this sailor's curse – in his case with fresh limes, prompting the American nickname for British sailors, 'limeys'.) Notwithstanding their fearsome names, *Erebus* and *Terror* also provided every possible comfort to while away the long dark winters. They carried extensive libraries including, as if to confirm what a home from home these vessels would be, at least one copy of that Victorian best-seller *The Vicar of Wakefield*. Confidence in making the passage was so high, crew members asked for mail to be forwarded to the Russian Pacific port of Petropavlovsk.

Franklin's orders were to enter the archipelago at Baffin's Lancaster Sound, and then when he reached Barrow Strait, 'to penetrate to the southward and westward in a course as direct as possible towards Bering Strait'. Only if this approach was blocked by multiyear ice, should he consider turning north.

As it happened, the main channel through Barrow Strait *was* blocked, so after a brief detour, the two ships anchored behind a small island for their first winter. The following summer (as we now know, but the subsequent search parties did not) they headed south along the western shore of the Boothia Peninsula, abreast James Clark Ross's Magnetic Pole marker, to the northern tip of King William Island – except that their unfinished charts did not show it as an island, because the younger Ross could not tell that from what he saw of its eastern coast on a sledging trip 16 years earlier. His uncle had therefore wrongly assumed it was connected to the mainland, and recorded it as such.

Ice trap

Franklin, not realising he had a choice, naturally sailed down the western side of the island, straight into a fatal trap. This was Victoria Strait, open to a mass of ice driving down another channel from the north-west. His ships were beset in the heavy offshore pack, and soon there was no way back. He had no radio to report his predicament. No one – except passing groups of Eskimos – knew where he was.

The Admiralty waited three years (the period for which *Erebus* and *Terror* had provisions) before sending out the first of a dozen or more search parties that were despatched over the next decade, and offering a £20,000 reward to tempt arctic whalers who might have useful information. Several of these rescue expeditions were personally organised by Lady Franklin, using ships she purchased. She even appealed to the president of the USA, where Congress voted funds to fit out two American vessels that joined the search. Her initiative was important in another way, because whereas a number of the naval experts thought Franklin must have headed north from the main Parry Channel, she rightly guessed he would search for a southern route – part of which he had, after all, mapped himself from the land some years earlier.

The first hard evidence of Franklin's fate was found in 1854 by Dr. John Rae, another of those Hudson Bay Company officials who seemed able to trek effortlessly for many hundreds of miles. Even the Eskimos admired the speed at which he could travel across such difficult terrain, and it was from them he heard about Europeans seen dragging sledges and a boat southwards along the western coast of King William Island because, as they indicated with sign language, their ships had been crushed by ice. Far more disturbingly, another group of Eskimos had found a camp site littered with corpses, with signs that the starving men had resorted to cannibalism.

No English gentleman...

When this shocking news reached London, Lady Franklin would have none of it. Charles Dickens protested indignantly that no English gentleman would descend to such depravity. But the Eskimos' account was confirmed by later investigation. In fact the evidence forcefully presented by naval historian Andrew Lambert's recent study of Franklin indicates not just isolated instances of cannibalism, but some wholesale butchery.

In any case, after Rae's news, the extent of the disaster could no longer be doubted. He even purchased one of Sir John's medals from the Eskimos, the gaudy emblem of a Knight Commander of the Royal Hanoverian Guelphic Order (now in the National Maritime Museum at Greenwich), of which he must have been immensely proud because he wore it round his neck for the blurred daguerreotype photograph taken just before he set sail.

In the course of his search, Dr Rae had mapped some of the last gaps in a tortuous route Amundsen was finally to negotiate half a century later. And at around the same time, two naval vessels, starting their search from the Pacific under the command of Captain Richard Collinson and Commander Robert McClure, discovered a deeper passage which linked up with Parry's great exploratory voyage of 1819.

Both men independently followed the Alaskan coast before striking north-east through Prince of Wales Strait until it was clear that, but for the ice, they could cross to Parry's 'Winter Harbour' or complete the North-West Passage by sailing east along Barrow Strait to the open sea. McClure backed off to have another try through the northerly strait that now bears his name – only to be permanently trapped by heavy ice and eventually forced to abandon his ship. Collinson also retreated, but then turned eastward along the mainland coast to demonstrate that this shallow, awkward passage linked up with the ice-laden waters in which *Erebus* and *Terror* had been crushed. Both men had discovered 'a North-West Passage' but had not conclusively resolved the Franklin mystery – or at least not to Lady Jane's satisfaction.

'My mission on earth'

With a savage war erupting in the Crimea, the Admiralty now closed the investigation and in March 1854 the names of Sir John Franklin and his fellow officers were removed from the Navy Lists, presumed dead. His wife's appeal for a 'final and exhausting search' was rejected. But she remained passionately concerned to find out exactly how he died, to recover the expedition's journals and thereby to establish his permanent fame as discoverer of the North-West Passage. For years after her husband disappeared, she wrote letters to him, hoping he would one day read them, and assuring him that the search was 'my mission on earth, it keeps me alive'.

Now she turned for help to Leopold McClintock and a crew of volunteers to man the steam yacht *Fox*, purchased and fitted out in 1857 with what was left of her fortune after funding previous independent expeditions. McClintock had hard-won experience from earlier voyages during which he became the navy's acknowledged expert in arctic sledging – a great asset in the kind of search he now had to undertake.

The voyage started badly, with the slender yacht trapped and almost crushed by ice before she left Baffin Bay. But once into the archipelago, McClintock worked his way as far south as possible – towards the area always favoured by Jane Franklin – and organised a systematic sledging survey around King William Island, consulting any Eskimos he came across.

Two skeletons, some chocolate and a little tea

The search ended on the island's west coast, where Franklin's despairing crews had abandoned much of their equipment before struggling towards the 'Back Fish River' they evidently knew was somewhere to the south. A note they left indicated that Sir John had died back in June 1847, probably before realising quite how hopeless the situation was. And on the same bleak shore, McClintock finally came across evidence that could not be denied – a sledge carrying a heavy naval boat with two skeletons inside, some chocolate and a little tea.

McClintock returned to a hero's welcome, a knighthood and the gratitude of Lady Franklin. She promptly set about ensuring that her husband's venture would be seen by the nation as a triumph not a disaster – since as far as she was concerned it was Sir John and the members of his expedition who should be credited with the discovery of the North-West Passage.

This claim – disputed by some – is permanently embossed on the bronze plaques surrounding the memorial statue erected in London in 1866 'by unanimous vote of Parliament'. Shaded now by tall plane trees, it stands in Waterloo Place, alongside the Athenaeum at the bottom of Lower Regent Street. Sir John (looking, I must say, rather more handsome than in his daguerreotype) gazes across to a statue of Antarctic explorer Captain Scott, with Florence Nightingale just up the road to his left. The main inscription reads: 'To the great arctic navigator and his brave companions who sacrificed their lives in completing the discovery of the North-West Passage 1847–8'.

In his guide to London's statues, John Blackwood suggests that of all its nineteenth-century monuments, ordinary people seem to have found this the most interesting. 'Crowds of ragged working folk' would apparently gather round the pedestal to hear one of their number recount the story of 'Lord' Franklin's tragic death. An intriguing piece of social commentary, since this was one of London's smartest residential areas; Gladstone and the Prince Regent were both locals in their time.

On either side of the statue's plinth are the borrowed words of the explorer's friend John Richardson – 'They forged the last link with their lives'. The suggestion is clear: that as the starving, scurvy-ridden remnants of Franklin's expedition trudged southwards along the west coast of King William Island, across Simpson's strait towards Back's river and the nearby 'Starvation Cove' (where human bones were later found), some must have realised they were at last skirting a passage – albeit not fully charted – which led through that frozen maze. By then, one imagines, none of them cared.

Even Franklin, who died much earlier, will have known from his navigational observations that he was not far from the 'Point Turnagain' he reached on one of his land expeditions. Yet the mass of multiyear ice in which he was trapped prevented him from demonstrating that his ships could close the gap, and he did not know that an alternative channel round King William Island lay to the east.

In truth the 'discovery' of the North-West Passage was a composite effort by Franklin's team and those that came in search of them – driven on by Jane Franklin's devoted persistence. It was left to the Norwegian Roald Amundsen to make the first complete transit under sail.

Going native

Setting out in 1903 in the 70-foot fishing smack *Gjoa*, it took Amundsen three years. He followed Franklin's track but for one crucial difference – when he came to King William Island, he took the easterly channel (the one John Ross decided was just a bay) which is protected from ice driving down from the north-west. Even then he only just made it through, running aground on a rocky reef at high tide, before wintering at the southern end of the island in the company of local Eskimos. He used the time, among other things, to calculate that the Magnetic Pole had shifted 40 miles to the north-east since James Clark Ross plotted its position in 1831.

As the *Gjoa* approached the Beaufort Sea the following summer, Amundsen was delighted to encounter the American whaler *Charles Hanson* out of San Francisco – a reminder, incidentally, that throughout a century of exploration, the whaling fleets were always there or thereabouts. They were usually too busy with their gruesome trade to be concerned about whether a through passage existed, but an exceptional captain like the Yorkshireman William Scoresby managed to combine slaughtering wales with extensive scientific research and, as we have seen, was influential in persuading the Admiralty to take up the challenge of the North-West Passage.

Although Amundsen revered Franklin and his men, he represented a fundamentally different approach to the practicalities of polar exploration. He adopted the Eskimo fur clothing – the *annuraaq* that gives us our word 'anorak' – and he was more than happy to make use of dog sledges, as he later demonstrated to devastating effect in his race with Captain Robert Scott to reach the South Pole. While the Norwegian dashed ahead, Scott continued to believe that 'no journey made with dogs can approach the height of that fine conception when a party of men go forth ... by their own unaided efforts'. Even today Englishmen can still be found carrying this often self-inflicted burden, trudging across the ice hauling a sledge.

Thirty years before Amundsen's successful passage through the Canadian archipelago, another Scandinavian, the Swedish baron Nordenskiold, had completed the North-East Passage along the opposite shore of the Arctic Ocean. His wealthy backers, including the Russian industrialist Sibiryakov, provided him with a more substantial vessel than the little wooden *Gjoa*. The 300 ton *Vega* was equipped with both steam engines and sails, and carried a steam launch for sounding ahead through shallow waters.

She set out from the Barents Sea in 1878 in company with three small freighters bound for the rivers Yenesei and Lena (in itself a pioneering effort). Progress was excellent along the relatively ice-free inshore passage, into the Laptev Sea and past the North Cape charted by James Cook exactly a century before. But there, at the end of September, only 120 miles from the Bering Strait, the sea was quietened by newly forming ice, the *Vega* was trapped and her 30-man crew faced another of those unimaginably long winters the early explorers endured. The ship was not released until the following July.

Nordenskiold had proved it could be done. That is why he was made a baron. But in his considered opinion, and given that he failed to get through in one season, the North-East Passage – or at any rate the far eastern section – was not going to provide a commercially useful route.

Short Cuts

Among the seven wonders of the ancient world, two were created in Egypt – the pyramids and the great lighthouse on the island of Pharos at the entrance to Alexandria. My personal updated list would include another Egyptian wonder, the Suez Canal. Not just because my first sight of a vast ship majestically crossing the desert skyline was so astonishing, but because of the way in which the canal transformed the shape of the trading world.

Comparisons with the changes in prospect if the Arctic Ocean continues to melt are not of course direct, not least because the northern routes will develop only gradually. But the Suez story (and that of the canal's Panamanian counterpart) does show how a simple geographic shift eventually impacts on every aspect of our lives, from the wars our young soldiers fight to the kinds of food that stock our kitchen shelves.

When the desert waterway opened on November 17, 1869 – which might also have been the opening night of Verdi's Egyptian opera *Aida*, had it been completed in time – Mediterranean ports were suddenly many thousands of miles nearer the Persian Gulf. The length of a voyage from London to India was reduced by about 5,000 miles, equivalent to several weeks. Australia seemed no longer quite so dauntingly remote, even if the actual distance saving was negligible.

By shortening busy trade routes, the canal also helped change the form of sea transport, shifting the balance between steam and sail. As William Bernstein explains in *A Splendid Exchange*, his splendidly illuminating study of world trade, iron steamships only gradually supplanted wooden sailing vessels during the nineteenth century because the newcomers were more expensive to build and operate – especially on longer voyages. The coal they needed occupied valuable cargo space, and their iron plates were more quickly fouled with weed and barnacles than copper-sheathed wooden planking, drastically reducing their speed.

So while the short-sea European trades quickly switched to steam, for a time the sailing clippers maintained their hold on the long haul routes. Then came the canal. It was always going to be a problem for sailing vessels, what with needing to be towed through, and persistent northerly winds in the Gulf of Suez. But simply by shortening so many of the routes to India and South East Asia, Bernstein points out, it jolted the economic balance sharply in favour of steam.

The canal's powerful geographical logic was already apparent to the ancient Egyptians, as an adjunct to the mighty Nile waterway. One of the Pharaohs seems to have tried to break through, and the invading Persian king Darius certainly created a working canal which survived through the later Ptolemaic dynasty before falling into disrepair. Unlike the modern Suez Canal, these early waterways linked the Red Sea to the Great Bitter Lake, then cut across to the Nile delta rather than driving northwards straight to the Mediterranean.

Napoleon's engineers found traces of the ancient works (the French, incidentally, having triumphantly dug their own short cut between the Mediterranean and the Bay of Biscay, the Canal du Midi, back in the seventeenth century), but a crucial surveying error persuaded them to abandon any attempt to recreate a desert waterway. They wrongly calculated that there was a difference of 10 metres (more than 30 ft) between the level of the Red Sea and the Mediterranean, which would have posed major problems.

In fact there is hardly any difference in sea levels, a remarkable fact enabling a 100-mile canal to operate without locks.

First night

It was the French diplomat Ferdinand de Lesseps who took on the construction task half a century later, against a promise from Egypt's Said Pasha that his company could run the new canal for at least 99 years. Work started in 1859, using forced Egyptian labour. Ten years later, French, British and Russian royalty were invited to the lavish opening ceremony. Verdi was apparently asked to compose something for the occasion, but in the event, his *Aida* was given its first performance at the Cairo opera house a couple of years later.

The British, predominant among the canal's early customers, made their first move to take control in 1875 when Prime Minister Disraeli secretly borrowed £4 million from the Rothschilds on the government's behalf – bypassing the Bank of England and without consulting Parliament – to buy out the Egyptian ruler's substantial shareholding. It was 'a political transaction', he later told MPs, to secure 'a highway to our Indian empire'. In 1888, Britain became the guarantor of the international waterway's neutral status.

Disraeli had declined to spell out for Parliament how his government would deal with the canal in time of war, but for the best part of a century it remained a focus of British imperial and naval strategy. In the Second World War, the 'Desert Rats' fought to halt Rommel's advance on Egypt. In 1956, when Colonel Nasser nationalised the Suez Canal Company, the British Prime Minister Anthony Eden led Britain into a disastrous military entanglement with France and Israel in a vain effort to regain control of the waterway. Later, following the 'Six-Day War' of 1967, it became the front line between Egyptian and Israeli forces, dug in on opposite banks during the so-called 'war of attrition'.

Canalside view

My personal memories of that conflict are vivid because as a journalist I reported from both sides – visiting an Israeli bunker to be mortared and bombed by the Egyptians (who must have noticed an unusual vehicle arriving, probably bringing people out into the open), only to find myself a few months later, somewhat disconcerted, peering out at exactly the same bunker from the opposite bank of the canal. I did not point out the coincidence to my Egyptian hosts.

More pleasant work was reporting on the eventual reopening of this wonderfully incongruous waterway in 1975, and recording its successful development. The canal company's major task was to cope with the rapid growth in size of tankers plying between Europe, North America and the oilfields of the Middle East. Excavated to a depth of 30 feet when it opened, the canal is now more than twice as deep, so that even the biggest vessels can pass through, at least in ballast. Ships normally proceed in convoys, at a stately maximum of eight knots to avoid damaging the banks.

The distance saved by using the canal obviously varies enormously, depending on the route. As an extreme example, a voyage between the Black Sea and the Red Sea is reduced by about 10,000 miles, an enormous saving in fuel and other operating costs. The sea passage from London to Mumbai is almost halved. Between North-West Europe and the Far East, the distance is cut by more than a fifth – not so different from the distance by which some voyages via Suez will in turn be reduced if the North-East Passage across the Arctic is fully opened.

The Suez Canal Company's commercial challenge is to set tolls which reflect the financial savings involved without squeezing the market too hard. At the time of writing, that calculation is complicated by both a worldwide trade recession and the persistent threat of piracy at the entrance to the Red Sea, which has prompted ship owners to route vessels round the Cape rather than pay additional insurance and risk capture. In 2009, the average toll for a single transit was about $250,000, but a big container vessel would pay much more than that.

The Panama Canal came later, inspired by the success of Suez. The French were again the instigators, but this time the nation which soon took control for strategic reasons was the USA.

Second time unlucky

The benefits of a canal that would enable ships to avoid the notorious dangers of the long haul round Cape Horn had been obvious for many years before it became practicable to build it – though a railway serving something of the same purpose was opened in 1855. But a decade after finishing his work in Egypt, Ferdinand de Lesseps began digging what he hoped would be another waterway that did not require locks. This time his calculations were wrong, his planning inadequate. Nor did he anticipate the devastating effects of malaria and yellow fever, which accounted for most of the 22,000 construction workers who died before the French effort was finally aborted in 1893.

The Americans took up the challenge at the turn of the century. Thousands more died, but work on this new maritime short cut was completed ahead of schedule just before the outbreak of the First World War. The prosaically named steamer *Ancon*, which had been running supplies for the construction, made the first ceremonial transit on August 15, 1914. The odd thing is that because of the way the Isthmus of Panama twists and turns between the two Americas, her passage from the Atlantic to the Pacific at that point was made in an easterly, not a westerly, direction. The 50-mile waterway makes use of the Gatun Lake on the Atlantic side, with vast sea locks at both ends.

Had the *Ancon* been on a working voyage from New York to San Francisco, she would now have had to steam just 6,000 miles, a saving of 8,000 miles on the voyage around Cape Horn. Ship owners and traders right across the world began weighing the benefits against the expected tolls. The economic balance shifted against the American transcontinental railway. US admirals pondered the limiting dimensions of the locks, which were not quite as large as they had originally hoped.

Given the canal's eventual success, sooner or later the Panamanians were bound to want to take control of it. The political argument took a while to develop, with negotiations for the handover beginning in the 1970s. The Americans were now cast in the role the British had played at Suez in the 1950s, although they managed to avoid starting a war, and of course they did not in any case see themselves as wicked imperialists. Panamanians proved no more incapable of running an efficient operation than Egyptians.

Over time, however, ships have outgrown the canal, with vast new container ships leading the way. Those built to the maximum dimensions are known in the trade as Panamax vessels. For example, a ship carrying 4,400 standard containers can just squeeze through. But ships of two or three times that capacity are now afloat.

The Panama Canal Authority has been forced to respond or risk losing some of its traffic to competitive routes – which may one day include the north-east and north-west arctic passages. In 2009, with its eye particularly on container traffic between China and the eastern seaboard of the USA, the Authority announced the start of a $6 billion enlargement programme to cater for the 'post-Panamax' era. By 2014 new and bigger locks, doubling the canal's traffic potential, should be ready to open their gates.

BOLSHEVIKS IN COLD WATERS

Fifteen years after Lenin's Bolsheviks seized power in St. Petersburg, the Soviet government took a strategic decision intended to demonstrate that Baron Nordenskiold was wrong, back in 1879, to dismiss the commercial potential of the Northern Sea Route. Except that by now the baron would hardly have recognised this exercise in state-directed economic and political development – potentially also beneficial to a geographically divided navy – as a 'commercial' venture.

The Soviet initiative was prompted in 1932 by the heroic success of the icebreaker *Sibiryakov* in completing the first transit of the Northern Sea Route in a single season – something Nordenskiold could not manage. It was considered 'heroic' because of the difficulties encountered along the way. Steaming east across the Kara Sea, the icebreaker found the narrow Vilkitsky Strait between the mainland and Severnaya Zemlya blocked by heavy ice she could not negotiate, which meant a long detour to the north. Then right at the end of the passage the *Sibiryakov* actually lost her propeller in the ice, only escaping from its grip by improvising sails from the canvas hatch covers.

That December the Kremlin responded with a decree establishing a major government department, the Chief Administration of the Northern Sea Route (abbreviated by Russians to Glavsevmorput) to develop, as part of a broader industrial programme, a practicable seaway from the White Sea to the Bering Strait. This meant building up a fleet of icebreakers, charting the often shallow coastal

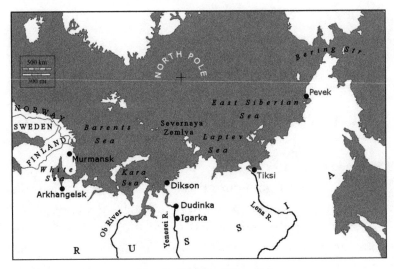

Map 7. Siberian Waters.

waters as opposed to simply mapping the coastline, establishing a network of meteorological stations to provide weather forecasts and, increasingly, providing aircraft to report on rapidly changing ice conditions.

The new organisation made an awkward start. In 1933, the brand new Danish-built steamship *Chelyuskin* – reinforced for arctic work but not an icebreaker – was crushed and sunk while attempting a second single-season passage from Murmansk to Vladivostok. But the Soviet authorities managed to present the crew's rescue from the ice as a triumph rather than a disaster (pilots who took part in the rescue were made 'Heroes of the Soviet Union'), and indeed the voyage is still celebrated today. When the wreck was located in 2006, only 150 feet down, there was talk, supported by Artur Chilingarov, of raising it to become a museum.

During the following seasons, the ex-Canadian icebreaker *Fyodor Litke* managed to complete the route from east to west and then lead a couple of freighters back. In 1936, she was in action again, this time escorting a pair of naval destroyers from the Baltic to the Pacific fleets, and dozens of merchant vessels were by now moving to and fro across some part of the arctic seaway.

Among the first solid trades to come under the Glavsevmorput umbrella (as opposed to supply convoys for outlying settlements or shipments of wretched prisoners to labour in the northern *gulag*) was timber from Igarka. In 1929, a saw mill was built at this tiny fishing village 400 miles from the mouth of the mighty river Yenesei, and within a decade, it had mushroomed into a logging boom town. It was an extraordinary place to construct a 'port', subject to violent changes in water levels as the seasonal ice dispersed, but foreign vessels were soon making regular visits to load timber for Western Europe.

So, four years in and things were looking good for the new Soviet organisation. But the 1937 season proved disastrous. The ice was bad, the shipping plans over-ambitious. Of the 64 vessels which set out that summer, 26 were trapped in the ice through the following winter, during which one of them sank. To make matters worse, they included almost the entire icebreaking fleet, leaving only the venerable *Yermak* to start extricating them the following year. Worst of all, this disastrous piece of bad luck or mismanagement coincided with Stalin's great political purge, during which so many thousands became scapegoats for his regime's failures.

The arrests included key Glavsevmorput officials, accused of sabotage. The head of the organisation Otto Shmidt (who had led the *Sibiryakov* expedition) somehow survived but was demoted and replaced by Ivan Papanin, a sycophantically reliable party man as well as an experienced polar explorer. Papanin had made his name as commander of the pioneering North Pole drifting station. He and three companions were landed there on an ice floe by aircraft – a remarkable feat in itself for those days – and left for nine months to drift slowly towards Greenland, taking meteorological and other scientific observations as they went.

With Stalin's compliments

One of the vessels trapped in the winter of 1937 was the *Sedov*, a little steamer expected to do an icebreaker's job but not really up to it. That October she was beset in the Laptev Sea and drifted 2,000 miles across the ocean for the next three years before breaking free halfway between Spitsbergen and Greenland. By coincidence this was roughly what the great Norwegian explorer Nansen had done in the *Fram*, in his case quite deliberately, in the 1890s. The *Sedov* actually crossed the *Fram*'s track several times. This made her crew's scientific observations doubly valuable – drift measurements, temperature readings, depth soundings and water samples, organised partly to ward off the terrible boredom and loneliness of the long winter months – because they could be closely compared to Nansen's work.

As the *Sedov*'s predicament worsened, the plight of her crew caught the Russian public's imagination. Newspapers picked up the story. The radio broadcast their musical requests. An arctic radio station on Cape Chelyushkin challenged them to a game of chess (lasting a year). And the Soviet authorities, while making strenuous efforts to rescue ship and crew, realised that for propaganda purposes they had here an example of what Papanin called 'the unshakeable resolve of real Bolsheviks'. Stalin duly sent them a congratulatory telegram.

The professional fortitude of the crew, especially the 11 members who volunteered to spend a third winter aboard, rather than be lifted off by aircraft or icebreaker, was indeed remarkable. The ice might leave them at peace for a while, but it was always dangerously on the move. A crack would suddenly appear in a nearby ice floe, swallowing scientific equipment they had mounted 'ashore'. A hiss and a roar would announce the approach of a pressure ridge easily capable of crushing their ship.

The late Terence Armstrong, from the Scott Polar Research Institute at Cambridge, provided a full account of the drift based on the crew's published journals. These were admittedly written with their party bosses' scrutiny in mind, but they show how even in such extreme circumstances there was no escape from the absurdities of Soviet ideology. For example, election day for the Supreme Soviet was both meticulously observed and celebrated with a concert. Armstrong notes laconically that 100 per cent of the electorate voted (for this purpose Sedov was part of the Arkhangelsk constituency), and at the end of the concert, it was announced that 100 per cent of the votes had been cast for the Communist candidates.

Glavsevmorput had meanwhile spent much of 1938 clearing up the debris of the previous season before resuming normal operations. By 1940 Europe was at war, but thanks to the Molotov–Ribbentrop Pact, Germany was not yet declared an enemy of the USSR. Indeed as we have seen, this was the year in which the German merchant raider *Komet* made its remarkable transit of the Northern Sea Route – with the help of Russian icebreakers.

Wonderland

Once engaged in conflict with the Germans, Glavsevmorput's role was to facilitate the arrival of allied relief convoys, not just from the Atlantic into Murmansk or Arkhangelsk but also through the Bering Strait, and to fend off Kriegsmarine raids on local shipping in the Kara Sea. These were mainly conducted by U-boats, having sometimes lain in wait for long periods, half submerged and camouflaged, and they cost the Russians a number of freighters and a few small naval vessels. During the German operation 'Wunderland' in 1942 (a romanticised choice of code name contrasting with prosaic British practice) the heavy cruiser *Admiral Scheer* sank the icebreaker *Sibiryakov* and then bombarded the isolated coaling depot at Dikson, causing the Russians serious logistical problems as well as damaging ships moored there. One of the cruiser's accompanying U-boats also revealed the darker side of the arctic shipping operation by sinking an NKVD (secret police) convoy carrying hundreds of political prisoners.

After the war, Glavsevmorput was busy re-equipping the river fleets, reviving the timber trade and establishing a series of drifting ice stations, modelled on Papanin's pioneering effort and supplied by air. It was the research conducted by these stations that confirmed the existence of the Lomonosov Ridge, now the centre of so much argument over the ownership of the ocean bed. Important work, but their sponsoring department's status had nevertheless declined, and in 1964 it was dismantled, its functions redistributed to the Ministry of Merchant Marine and other authorities.

Yet this evidently did not reflect on the practical importance of the Northern Sea Route. New icebreakers were extending the length of the navigational season; the volume of cargo moving along the route was rapidly increasing. Then in 1967 the Russians surprised, indeed astonished, the international shipping community by issuing an open invitation to make use of it as an alternative to the southern oceans.

Wishful thinking

In March that year, 'after a careful study of Soviet experience', foreign vessels were offered the opportunity to save an estimated 13 days and about £15,000 in operating costs (at historic prices) on an illustrative voyage between London and Yokohama when compared with the Suez route. Icebreakers and ice reconnaissance aircraft would be at their disposal; they could make use of Soviet radio and meteorological services. There would be new charts, pilots if required, and advice on the operation of arctic convoys. It seemed that four centuries after European seamen started searching for it, the North-East Passage they dreamed of was about to become a reality.

Just what prompted the Russian invitation at this particular time – in contradiction to years of paranoid institutional secrecy – is difficult to discern. An East–West Cold War persisted following the Cuban missile crisis; military paranoia was at a painful pitch; the arctic islands of Novaya Zemlya straddling the Northern Sea Route were still being used to test nuclear bombs – although the dreadful era of atmospheric testing, which in Britain prompted the Campaign for Nuclear Disarmament (CND), had admittedly been ended by international agreement.

Maybe the Russians already calculated that simmering conflict in the Middle East – where they themselves were helping to stir the pot – would threaten the Suez Canal and increase the attraction of an alternative route. But the Six-Day War between Israel and her Arab neighbours, which did lead to an eight-year closure, was still a few months away.

My guess is that the Russian initiative had something to do with the Ministry of Merchant Marine's urgent drive to earn foreign exchange by selling shipping services of all kinds. The Ministry was headed by Viktor Bakayev, a long-serving Soviet official whose career in charge, from 1954 to 1969, roughly paralleled that of Admiral Gorshkov as head of the Soviet Navy. (Bakayev once joked that he had invited three Italian shipping ministers to visit him in Moscow but none of them lasted long enough to make the trip.)

'Strength through joy' for Stakhanovites

The merchant fleet Bakayev inherited was a motley collection of pre-war Russian vessels, a number of American 'Liberty' ships never returned after the Lend-Lease programme, plus whatever Soviet forces acquired from Finland and the annexed Baltic states, or demanded from Germany as reparations. For example, the Russians managed to salvage and refit no less than 15 German passenger liners. The Hamburg-built *Hansa* reappeared as the *Sovietsky Soyus* (Soviet Union); the *Marienburg*, built in Stettin at the outbreak of the war specifically to provide 'Strength through

Joy' cruising holidays for deserving workers of the Third Reich, ended up carrying worthy Communists who had over-fulfilled their production norm.

By comparison with other nations' fleets in the 1950s, Soviet vessels were in general old, small and under-powered. But they had to cope with the demands of a growing seaborne trade, heavy coastal traffic (including the arctic routes) and a variety of foreign aid programmes. Ten years on, hundreds of Soviet ships were delivering food, industrial equipment and armaments each year to North Vietnamese ports and hundreds more were 'discharging the Soviet Union's sacred internationalist duty by rendering fraternal disinterested assistance to the Cuban people'.

The Bakayev plan

Some of this traffic had to be handled by Soviet vessels. Elsewhere it might be possible to charter foreign tonnage, but that probably had to be paid for in precious hard currency. So for Bakayev, expansion of the Soviet-flag fleet was an imperative, not merely to stop the currency drain but if possible to turn its deep-sea operations into a net earner. By the end of his career, he had developed a sort of strategic 'Bakayev Plan', akin to Gorshkov's dream of worldwide sea power. Having first secured political and economic independence for the USSR's external trade, Soviet ships would improve the balance of payments by joining the cross trades, catering for foreign tourists, infiltrating Western shipping cartels – the so-called 'conferences' – and then eventually promoting a utopian world in which the UN would gradually take control.

Somewhere in this grand design was a corner for the Northern Sea Route. Bakayev's officials had even worked out the charges they would make for icebreaking, pilotage and meteorological services, on a sliding scale dependent on the extent to which ships were strengthened for work in ice. But hard-headed Western ship owners did not share the minister's vision. Only the Japanese showed much interest. Few foreign vessels were already strengthened for arctic work; there was some confusion, after all, about the service charges; insurance costs were bound to be high and the whole project had an air of unreality. The Russians own freighter *Novovoronezh* – ice-strengthened, naturally enough – loaded a demonstration cargo at several European ports and apparently reached Yokohama in 28 days. But no foreign vessel followed her. Moscow's invitation was not formally renewed for another 20 years, in a speech delivered in Murmansk by the then president Mikhail Gorbachev.

Arctic gateway

The port of Murmansk is pivotal in the story of the Northern Sea Route. It is the ice-free gateway between the North Atlantic and the Barents Sea, and the departure point, along with the White Sea port of Arkhangelsk, for much of the shipping activity stretching in measured stages to the furthest shores of Siberia. Heavy investment in expanded shoreside facilities is now promised, with a forecast cargo throughput of 100 million tonnes by 2015.

For all its modern importance, the city was established only in 1916, as the terminus of a new railway from St. Petersburg constructed to supply the war effort. Its name derives from the old Russian word for the Norwegians – the Murmans – and indeed for the Barents Sea.

The new port was established on a fjord penetrating the Kola Peninsula, which stretches from the White Sea in the east to the Norwegian border in the west (incorporating a sliver of territory that for a few years between the two world wars gave Finland access to the northern sea at Pechenga). Kola is nowadays extensively, grimly industrialised – by mining, fishing and, increasingly, various services to offshore oil operations in the Barents Sea. During the Cold War, it also acquired a vast military complex and is still strategically vital as the home of Russia's Northern Fleet, with its headquarters near Murmansk at Severomorsk.

More importantly in the present context, the city is home to the Murmansk Shipping Company (MSCO) – company logo a polar bear. With its subsidiaries, MSCO controls a large proportion of shipping along the Northern Sea Route, and until 2008, it also managed Atomflot, the state-owned fleet of nuclear-powered icebreakers which has been crucial to the route's development.

The company began as a state enterprise in 1939 providing bulk cargo and passenger services but following the collapse of the Communist regime became a joint stock company, along with most of the Russian shipping companies operating in the North. A combined fleet of 300 vessels, many of them specially strengthened against ice, handles bulk cargoes like grain, ores and coal (including shipments from Spitsbergen), general cargo in containers, timber and more recently oil.

The city of Murmansk is proud of its history, especially its desperate resistance to German attack from Finland in 1941 – commemorated by a huge statue of a Soviet soldier overlooking the port – and its role in the development of the Arctic. The MSCO is also proud to have 'added several glorious pages to the history of the Soviet and Russian navigation', many of them naturally involving the specialised business of icebreaking (of a dozen 'legendary' ships pictured on the company website, nine are icebreakers). For good or ill, therefore, Moscow's decision to remove the nuclear-powered icebreaking fleet from MSCO's management marks a significant change of direction.

Atomic relations

Nuclear technology is now an all too permanent part of our world, and almost from the start, this arctic region – by which I mean the Barents and Kara seas, and the great rivers flowing into them – played an active part in making it so. Most obviously, it was chosen for the testing of atomic bombs – on the islands of Novaya Zemlya. The nuclear-powered submarines and the icebreakers came later. But it is a dangerous characteristic of this technology that military and civil applications are closely related.

British and Russian plutonium factories provided the patterns for early power station designs (of which Chernobyl was a disastrous example); the Americans chose to adapt their submarine propulsion technology for electricity generation. And whatever path the engineering development took, peaceful or otherwise, it eventually produced toxic nuclear waste for which there is no completely safe disposal. For many years the Russians just dumped much of theirs around Novaya Zemlya, or drained it into the Ob and Yenesei rivers, flowing northwards to the Kara Sea.

Having exploded their first atomic bombs in Kazakhstan, the Russians began testing them over Novaya Zemlya in the 1950s (first removing the indigenous people dumped there in the nineteenth century to keep out the Norwegians). In those early days all tests, whether by the USA, Britain, France or the USSR, were conducted in the atmosphere, leaving the radioactive fallout to be swept round the globe until it drifted down, or was washed down by rain – just as much later the drifting cloud from the Chernobyl power station explosion settled on the Welsh hills.

Czar Bomba

The Soviet programme reached its awful climax in October 1961. The Cold War was at fever pitch, the Berlin Wall under construction, and the Soviet leader Nikita Krushchev wanted a show of strength. He ordered the detonation of the largest nuclear device ever built, the 50-megaton 'Czar Bomba' – given that playful nickname by analogy with 'Czar Kolokol', the 200 ton, 20-foot-high bronze bell tourists gaze at in the grounds of the Kremlin.

The bell was made in the eighteenth century for Anna, the niece of Peter the Great; it never rang because it cracked during the casting process. The awful Czar Bomba, by contrast, shook the earth like nothing before or since. It was heard and seen far across the Barents Sea in Scandinavia, even though its explosive yield had been deliberately reduced to limit contamination, and give the Tu-95 bomber crew who dropped it time to get clear.

It was after this exercise in military insanity that the physicist Andrei Sakharov began to emerge as a dissident. Indeed, antinuclear movements everywhere, including Britain's CND, were especially concerned at the indiscriminate effects of atmospheric testing wherever it occurred, and it was this pressure, added to some belated political common sense, which led to the Partial Test Ban Treaty of 1963 – after which most testing was done deep underground. On Novaya Zemlya, an underground programme began in 1964 and was finally halted in 1990.

Half-life

But for the Norwegians who share the Barents Sea with the Russians, this was not the end of the story. They remain concerned at the extent to which these otherwise clean waters have been permanently contaminated by the dumping, deliberately and accidentally, of nuclear reactors and spent fuel elements. The end of the Cold War, followed by Gorbachev's *glasnost,* has admittedly removed much of the secrecy and some of the fear. The USA, Norway, Sweden and Britain have provided Russia with financial and technical assistance in improving safety standards, particularly where nuclear waste management is concerned. But much of this pollution does not decline on a conveniently human timescale – the radioactive half-life of plutonium-239, the active ingredient of many nuclear processes devised by the military, is 24,000 years.

Apart from what may lie off the shores of Novaya Zemlya, there is concern about the radioactive waste which for many years flowed down the rivers Ob and Yenesei, emptying into the adjacent Kara Sea. Three of Russia's former 'secret cities', industrial complexes producing bomb-making material like plutonium, often using forced labour, are located on the head waters of these great rivers. And for decades these factories routinely dumped their waste in nearby lakes and streams (or sometimes stored it in steel tanks, which on a couple of occasions exploded). The local contamination was horrendous, as radioactive material was trapped in river sediments or spread across the flood plains, and a small amount eventually made the 1,000-mile journey to the arctic shore.

This is not to suggest that nuclear contamination is an immediate health risk in the Arctic, or that it threatens fish stocks and other wildlife in the same way as a big oil spill. Its current significance lies elsewhere, in that it will prompt anxious questions about future Russian attitudes to industrial safety and marine pollution as oil platforms appear in the Barents Sea and shipping proliferates along the Northern Sea Route.

'You get used to it'

In the past, for whatever reason, attitudes have been notoriously cavalier, especially in nuclear matters. A Russian defence minister was quoted as saying there was less radiation on Novaya Zemlya than in his office – a cheerful insouciance I have heard before. 'We don't worry about radiation', a worker at the devastated Chernobyl nuclear power station told me as he showed me round, 'you get used to it.'

The Russian authorities responsible for arctic development nowadays make a point of responding positively to concern about environmental damage, whether it comes from their own activists or from the international community. But the hard fact remains that offshore oil operations can never be free from the risk of pollution – much more difficult to deal with in icy waters – and another worry, at least for foreign ship owners, is the prospect of manoeuvring at close quarters with vessels containing a nuclear reactor.

In March 2009, for example, the Russian salvage coordinating centre reported a collision in the Kara Sea between the nuclear icebreaker *Yamal* and the 16,000 ton tanker *Indiga* – one of the vessels which shuttles between the coastal oilfields and a floating storage depot in Kola Bay. The heavily built icebreaker suffered no damage, but the tanker was reportedly left with a 30-foot crack in her main deck and had to be escorted back to Arkhangelsk for repair. Had she been full of oil at the time, the story might not have had such a happy ending.

In 2009, the Scandinavian environmental watchdog Bellona raised the alarm at Russian plans to build floating nuclear power stations for offshore drilling in the Barents and Kara seas, or to support remote Siberian communities. These self-propelled platforms are expected to store their own waste, but again, unfair though it may now be, the earlier Soviet record does not reassure.

Breaking the Ice

With all due respect to its remarkable – often indispensable – qualities, the icebreaker is something of a maritime freak. On the open sea, with its strange sawn-off bow, top-heavy superstructure, a tendency to roll and slam, few would consider it an ideal of sea-going beauty. At work in heavy ice, it is inevitably uncomfortable as it backs and charges and crunches its way through. Yet it must be a wonderful sight from the deck of a vessel trapped in ice, awaiting rescue. And without its help, current operations in the Arctic would be impossible.

Americans lay claim to the first steam-powered icebreaker, Philadelphia's *City Ice Boat No. 1*, built to clear the harbour in 1837. The first recorded Russian vessel was the converted tug *Pailot*, used by a merchant in 1864 to clear a passage across the frozen River Neva between St. Petersburg and Kronshtat. But the first true sea-going vessel appeared right at the end of the century – the *Yermak*, named after the Cossack leader who explored Siberia, and sometimes referred to by Russians as the 'grandfather of icebreakers'.

The enthusiastic Admiral Makarov persuaded the Czar's government to have her built at Armstrong-Whitworth's Newcastle-upon-Tyne yard – a bulbous 5,000 tonner with tall smokestacks and steam-reciprocating engines delivering 10,000 horsepower. The British shipbuilders did an excellent job; she was still in service more than 60 years later. But the admiral seems to have oversold his project by painting a vivid picture of her carving a path straight to the North Pole, and when she proved not quite up to it, his more practical plans for opening up the Kara Sea were unnecessarily discredited.

It took the debacle of Tsushima to persuade the Imperial regime to build another pair of icebreakers to explore the North-East Passage. They set out in 1913 under Commander Vilkitsky, after whom the crux of the route, the strait between the Kara and Laptev seas, was subsequently named.

As we have seen, the *Sibiryakov*'s achievement in just about completing the Northern Sea Route in 1932 was instrumental in persuading Stalin to promote its development, and a few years later, the first of a class of Russian-built icebreakers – still old-fashioned coal burners, but effective ships nonetheless – was named after the Soviet leader. After the Second World War, the Soviet regime continued to give the arctic route a high priority, even offering in 1967 to open it as an international seaway. Then over several decades a dozen powerful diesel-electric icebreakers were ordered from Finland, while the Russians themselves embarked on a pioneering effort to apply nuclear propulsion to this type of ship.

Today, a dozen countries operate icebreakers. Canada needs them in large numbers, including a couple of heavy vessels, to cope with winter in the St. Lawrence and Hudson Bay, as well as the Arctic; Scandinavians use them to keep Baltic ports clear, with Finnish shipyards specialising in their design and construction. The USA has strategic and scientific interests in both Antarctica and the Arctic, for which it has three polar-class vessels.

But no one disputes Russia's predominant role at the heavy end of this business, or its unique experience in adapting nuclear propulsion to icebreaking, even if they question the economic logic behind it. Other nations did experiment with different kinds of nuclear merchant ship in the 1960s and 70s – the Americans built the *Savannah*, which was at least technically successful, and Germany's *Otto Hahn* was in operation for about ten years. But for a mixture of economic and political reasons, nothing finally came of these ventures. For one thing, they ran into a wave of public anxiety about nuclear power in the West, followed by an alarming spate of ship collisions that led to the introduction of mandatory shipping lanes.

Russian engineers, on the other hand, had no such inhibitions, and in the icebreaker they found an ideal vehicle for this form of propulsion. Such a vessel needed a great deal of power and the ability sometimes to remain at sea for long periods without refuelling – both things a nuclear reactor could deliver.

A nuclear pioneer

Their first effort was launched in 1957 and named after Lenin (a name thankfully not susceptible to the political changes that afflicted others). She was a 16,000 tonner with three reactors producing 44,000 shaft horsepower to drive triple propellers. Her endurance was practically unlimited.

The *Lenin* was not only the world's first nuclear surface ship (just beating the *Savannah* into service), she was also the first to do really useful work as well as being a floating test bed. She could plough ahead at three knots through 1.5 metres (5 ft) of ice and, if necessary, smash through 3–4 metres (10–13 ft). In her first five years, she steamed 50,000 miles, escorting 400 vessels –'steamed' being the operative word, since her reactors heated water to provide steam for turbines coupled to electric motors driving the propellers. It became clear that year-round operations were feasible on at least part of the Northern Sea Route. With studied nonchalance, Soviet officials declared that 'she could go to the North Pole, but there is no reason to divert her from her duty of convoying ships merely to break one more record'.

Her operation was far from trouble-free, however. During refuelling operations in 1965, a reactor core apparently overheated, distorting the fuel elements. Then a couple of years later the cooling system reportedly sprang a leak, and the difficult, dangerous work of finding it did irreparable damage. But since this was the Soviet era, such problems were kept as secret as possible. No public explanation was offered. As far as Western observers like me were concerned, in 1967 the *Lenin* just disappeared.

There were inevitably some wild rumours: that a crane driver had wrecked the ship by dropping a heavy chunk of reactor, or Russian sailors suffering from radiation burns had abandoned her in the ice. But a more prosaic hypothesis proved correct – that her three reactors were being replaced by a more compact twin-reactor system, test bed for a whole new class of even more powerful vessels.

So what makes a successful icebreaker? One might think the answer would be a ship designed with a sharp, specially strengthened bow to cut a way through. And in fact a few vessels have embodied this 'ice-cutting' principle, notably the slender clipper-bowed *Fyodor Litke*, launched in 1909 and looking more like an Edwardian millionaire's steam yacht than the hard-working ship she proved to be.

Like the *Yermak*, she was built in a British shipyard – at Barrow-in-Furness – but in her case for the Canadian Gulf of St. Lawrence. Acquired by Russia at the beginning of the First World War, she fell into Bolshevik hands and for the next 40 years was involved in every adventure the Soviet Arctic had to offer – leading convoys, rescuing explorers, escorting warships, plus the shameful work of servicing Siberian gold mines.

Out of the ordinary

However, the basic method used by nearly all icebreakers is not to cut a way through, but use the ship's weight to break the ice from above. Their characteristic profile, therefore, has a sawn-off bow to ride up on the ice. The double-skinned hull is smoothly rounded – which is why these vessels tend to roll – to reduce drag and disperse the broken ice. A broad band of plating along the waterline and around the bow must be much thicker than in a normal ship to withstand the abrasive power of sheet ice, with extra internal stiffening to prevent the hull being crushed. The rudder has to be especially strong and as far as possible protected from ice damage when going astern, because icebreakers often back and fill to clear water for another vessel or break through a ridge. Propellers are also shielded and strengthened, typically with some sort of access hatch so that a damaged blade can be inspected and if necessary changed at sea. Modern icebreakers also use underwater air bubbles to 'lubricate' the hull and a technique of deliberate rolling – by shifting ballast water around fast – to shake themselves clear of ice. Altogether an extremely complicated and expensive piece of equipment.

Add to all these special design features a pair of nuclear reactors delivering 75,000 horsepower through a linkage of turbines and electric motors and you have the *Arktika*, first of a class of six immensely capable Russian ships. She was launched in 1975 and two years later became the first surface ship to reach the North Pole, where the mean thickness of the ice is about 3–4 metres (10–13 ft). She once spent a year at sea without putting into port – another advantage of atomic power – and her reactors were not finally shut down until 2008.

Disposing of such a ship is not a simple matter, however. Before she can safely be scrapped, radioactive fuel elements must be removed, the dangerous nuclear waste stored and the irradiated reactor core dismantled. This complex decommissioning process requires an elaborate supporting infrastructure whose expense and environmental risks should be added to the other costs of providing the nuclear fleet's unique capability. But one suspects that, as with the early British nuclear power stations, this part of the cycle was never realistically planned for, let alone fully costed. It is certainly not complete.

6. Nuclear icebreaker *Rossia* followed by freighter *Beluga Foresight* on Northern Sea Route, September 2009.

Back in the 1970s, the Russian authorities were more concerned to emphasise the positive prospect of extending the arctic navigational season and opening up the Northern Sea Route as a whole. In the spring of 1978, the *Arktika*'s new sister ship *Sibir* followed the previous year's polar triumph by escorting an ice-capable freighter right across the Soviet Arctic, ignoring the inshore straits to pass north of all the main Siberian islands.

A difficult birth

Six *Arktika*-class vessels were eventually built. The last of them – the *Ural* – was launched from the Baltic Yard at St. Petersburg in 1993, just as the Communist system was disintegrating to leave the Russian economy slumped in a chaotic process of capitalist reform. The funding dried up. She was not completed for another 14 years, finally emerging in 2007 under a new name – *50 Let Pobedy* ('50 Years of

Victory') – not one that trips easily off the tongue when radioing from a ship's bridge, whatever language you choose. In keeping with modern concern for the environment, she is 9 metres longer than her sisters to make room for a waste-processing unit. Another sign of the times is her purpose-built accommodation for more than 100 passengers, enabling her to join the lucrative new business (at least in foreign currency terms) of taking tourists on cruises to the North Pole.

With Finland's help, the Russians had meanwhile pioneered two other applications of nuclear propulsion – a pair of shallow-draft icebreakers and a so-called LASH (lighter-aboard-ship), a sort of arctic mother ship that carries either barges or containers. Both designs reflect the need to integrate ship transport along the Northern Sea Route with small vessels and lighters that can penetrate its shallow estuaries and rivers, though in practice, it seems that offshore transhipment by a LASH carrier has proved difficult in bad weather, or just more trouble than it is worth.

The two specialised icebreakers, *Taimyr* and *Vaygach*, were designed and built in a Finnish yard, then towed to Leningrad for installation of their nuclear reactors – examples of a long-standing technical co-operation between the two countries. The Finns know about ice. Their own ports are ice-bound in the winter, and since the 1960s, they have supplied the Russians with a range of diesel-electric icebreakers for polar, river and harbour work. Several of the bigger vessels were specifically designed with a shallow draft (not normally a desirable characteristic in an icebreaker) to operate in Siberia's coastal channels and the estuaries of the Ob and the Yenesei.

Atomic takeover

Until 2008, Russia's nuclear-powered icebreakers were all managed on the state's behalf by the privatised Murmansk Shipping Company. Indeed, the exploits of this unique fleet – the record-breaking trips to the North Pole and so on – figured largely in the company's literature and were proudly exhibited in its museum. But that August, after recurrent arguments about pricing and funding, control of the operating company Atomflot was transferred to the state nuclear agency Rosatom – presumably on the basis that if the costly nuclear fleet was not commercially viable but nevertheless worth supporting for wider reasons, it made more sense for the central government authorities to control its management and decide how revenues were distributed.

Whatever the reasoning, the change was described locally as a 'painful process'. It called into question not just the future management structure but also the Kremlin's long-term strategy for the Northern Sea Route. Would it still have a place for expensive nuclear icebreakers or just build more of the conventionally powered vessels on which other arctic nations rely?

Some reassurance for those whose jobs were at stake came that October from a meeting, reported on the government website, between Prime Minister Vladimir Putin and the head of Rosatom, Sergei Kiriyenko. Putin talked about 'assessing the state of the fleet', adding that he had of course in mind 'our plans for the further development of the Northern Sea Route'. For his part, Kiriyenko warned that without new vessels to replace the ageing fleet, opportunities in the Arctic would decline after 2015. Building and funding them was therefore 'a pivotal task'.

There was an equally positive message from the Central Marine Research and Design Institute (CNIIMF) in St. Petersburg – professional cheerleaders for the Northern Sea Route. Its directors are still planning on the assumption, or at any rate the hope, that three new nuclear-powered variable-draft icebreakers will be available by 2020 (along with five smaller diesel-electric vessels). These powerful new ships – their variable draft another pioneering feature – should be able to cope with 3 metres (10 ft) of ice. They will replace five ageing nuclear vessels including the shallow-draft *Taimyr* and *Vaygach*, both scheduled to retire by 2013.

There is nevertheless concern that because of the usual funding difficulties this replacement programme, like others before it, may be allowed to slip. Since the first of the new class will not be commissioned until 2015, there will in any case be a two-year 'icebreaker gap', particularly affecting operations on river estuaries where the depth of water is a limitation.

A polar giant

Beyond that there is the promise – still at the conceptual stage – of an immensely powerful icebreaker capable of operating 'under any ice conditions in any area of the Arctic'. Such a creature could escort freighters straight across from the Barents Sea to the Bering Strait.

Atomflot's immediate commercial prospects have meanwhile been dealt a blow by one of its important customers, the vast Siberian mining and smelting combine Norilsk Nickel, which decided to build its own fleet of specialised freighters so as to be as far as possible independent of icebreaker support. The combine was established in the 1930s by the Soviet regime, initially using forced labour. Privatised in the 1990s, with major subsidiaries in the USA and Canada, it is now one of the world's largest producers of nickel (used in stainless steel), palladium, platinum and copper, plus by-products such as cobalt, silver and gold. There is grim irony in the fact that platinum and palladium are active components of catalytic converters – expensive devices to reduce toxic emissions from our cars – yet the notorious Norilsk smelters pump out vast quantities of poisonous sulphur dioxide, threatening local people's health and causing acid rain over a wide area.

Supplies are shipped into Norilsk, and metal products shipped out, through the nearby port of Dudinka, on the River Yenesei. The commercial importance of this trade can be judged from the fact that in 1978 Murmansk–Dudinka became the first Russian arctic shipping route to be operated on a year-round basis. Moreover, the mining company's decision to run its own fleet is doubly significant because the five new freighters are not merely ice-hardened, they are so-called 'double-acting ships' which do their own icebreaking.

Back to front

A double-acting ship tries to have the best of two different worlds – with a conventional bow shape that is efficient in open water and a stern designed for icebreaking. So it moves forward in a normal way until it is impeded by ice, then turns round and goes happily backwards, controlled from a secondary, backward-facing bridge – or so I am assured, though not having sailed on one of these extraordinary vessels, I still find the whole idea decidedly odd. The 'stern ahead mode' is achieved by swivelling an electrically driven propeller unit – an 'azipod' – through 180 degrees

7. The double-acting ship *Norilsky Nickel* – is this the future of arctic operations?

instead of reversing the propeller in the normal way (electric drive also helps smooth out the shocks delivered by blocks of ice swirling past). The result is a part-time icebreaker which retains much of a normal vessel's hydrodynamic efficiency.

The azipod itself has been around for a while, for example on cruise ships, but its newfound ability to 'eat' ice has the potential to change the whole balance of arctic shipping operations. It is already used by shuttle tankers in the Baltic and the Barents seas. For the metals combine, it was applied to a small double-acting container ship, the 18,000-ton *Norilsky Nickel*, by Finland's Aker Arctic Technology. The Finnish company also has a design for a liquefied gas variant on its drawing board that may one day operate in the Canadian archipelago and the Beaufort Sea, as well as in the Barents.

During trials in the Kara Sea in the winter of 2006, the *Norilsky Nickel* seems to have proved if anything more capable than expected, coping with deep ridges as well as the 1.5 metres (5 ft) of flat ice for which she was designed. She and her sister ships are now said to be doing well on the Dudinka run.

Sovcomflot, the big state-controlled tanker operator, has meanwhile acquired a series of ice-strengthened azipod tankers to shuttle between offshore terminals in the Barents Sea and Murmansk, where the oil is transhipped into larger vessels for export. Such ships are supported where necessary by the oil companies' own small icebreakers – another move towards operational independence.

The *Vasily Dinkov*, first of three 70,000 tonners described as 'multidirectional' rather than fully double-acting, was handed over by Samsung, the South Korean shipbuilders, at the end of 2007. She too was reportedly capable of driving through 5 feet of ice, and intended to work from Lukoil's Varandey terminal, 14 miles offshore in the cold south-eastern corner of the Barents Sea. Two similar ships to service the nearby Prirazlomnoye field were ordered by Sovcomflot from Russia's own Admiralty shipyards in St. Petersburg.

The initial construction cost of such vessels, with much reinforced hulls and diesel-electric propulsion at 16 knots through twin pods, is higher than that of a conventional tanker, but the economic benefits of independent operation are evidently expected to outweigh the construction cost. The next question is whether as the Arctic grows warmer, a fully double-acting ship can usefully be developed to dispense with icebreaker support on longer, more exposed routes where multiyear ice may be a problem.

One way or the other, the record ice-free season of 2007 will be seen in retrospect to mark a change of course for Russian icebreakers. The last of the older generation of nuclear ships, the *50 Let Pobedy*, eventually entered service in April that year. Eighteen months later, the class prototype, the *Arktika*, shut down her reactor after 33 years of operation. From then on, Atomflot faced the progressive retirement of its ageing fleet until 2020, when unless the promised replacements are funded and delivered, only one nuclear icebreaker will still be working.

Nuclear tourism

The old *Lenin*, revered as a Communist pioneer, marked her 50th year by embarking on a new career as a floating museum and tourist attraction. In the summer of 2009, she was moved temporarily to a specially adapted berth in Murmansk to test the arrangements for her new role, and it was hoped to have her permanently moored there to celebrate the anniversary of her entering service. Several thousand people visited the ship to be given hour-long conducted tours, including a view – through thick plate glass – of the stripped out engine room.

None of the other arctic nations have a coastline or maritime strategy that justifies icebreaking investment on the Russian scale, although in some respects Canada obviously comes close. But most of them – including Finland and Sweden in the Baltic – operate icebreaking fleets of various shapes and sizes to suit their local needs. And those that share the Arctic Ocean with Russia are watching the weather with an anxious eye.

For years now, the US Coast Guard has been lobbying, with the support of US scientific academies and elements of the Pentagon, for funds to strengthen the American presence in the Arctic by building new icebreakers – or at least refurbishing the existing ships. The Coast Guard currently has just three heavy (non-nuclear) icebreakers. The *Healy* is a modern vessel devoted mainly to the arctic research for which she is extensively equipped. The remaining pair, the elderly *Polar Star* and *Polar Sea*, have outlived their original 30-year design lives, and the former, commissioned back in 1976, is laid up in Seattle in so-called 'caretaker' status.

In recent years, the Americans have therefore been stretched to find one or more powerful icebreakers to enable supply ships to reach their Antarctic scientific research station in McMurdo Sound (on a couple of occasions they had to ask the Russian icebreaker *Krasin* to help out). And now they realise that at the other end of the world a changing climate will probably put greater demands on their small polar fleet by opening up the Arctic to maritime transport, oil exploration and tourism.

'Water where it didn't use to be'

A 2007 report from the US National Research Council warned that because of a lack of funds, the nation's icebreaking capability 'is today at risk of being unable to support national interests in the north and the south'. After surveying the Alaskan shore in 2008, the Coast Guard commandant Admiral Thad Allen expressed his concern more simply: 'All I know is, there is water where it didn't used to be, and I'm responsible for dealing with that.'

The core US national interests in the Arctic are retaining freedom of navigation, especially for naval forces, and the ability to conduct independent scientific research. For the moment, American commercial activity is not generating maritime traffic in the way Siberian industries do – and in any case US Coast Guard icebreakers are not servants of the oil industry.

When oil was discovered on the north slope of Alaska, it was thought initially that it might be brought out by sea, but the failure of the experimental icebreaking tanker *Manhattan*'s winter voyage across the North-West Passage in 1970 put paid to that, leading to a strategic commercial decision to bring the oil south through the Trans-Alaska Pipeline. The big Red Dog zinc mine in north-west Alaska is served by massive ore carriers, but they only load there during the few ice-free summer months.

This situation may well change, however, if calls for more drilling from the likes of former Alaskan governor Sarah Palin are answered. And scientific interest in the region is certainly not going to slacken. One way or the other, there will be more maritime activity right across the region, and sooner or later, that will involve emergencies in which a fully ice-capable ship is essential if the USA wants to play its part – a serious oil spill, for example, or an emergency involving one of the many cruise liners that now head north each summer to witness melting glaciers, stranded bears or migrating whales.

The National Research Council report for Congress recommended that the USA should 'continue to project an active and influential presence in the Arctic'. And the way to do that, it argued, was to build two new polar-class icebreakers as 'multimission' ships to support the *Healy*. These would not in any case enter service for another 8–10 years (when the Healy would be almost 20 years old), during which time the elderly *Polar Sea* would have to be kept in working order, with the *Polar Star* in reserve.

Underlying all this discussion of operational requirements is the sensitive issue of funding. For which of these 'multimissions' should the Coast Guard pick up the bill, and for which should it be handed on to the Pentagon, the National Science Foundation or some other department?

Canada has been conducting a similar debate about future requirements, made more complex by periodic spasms of public anxiety about the need to assert the nation's 'sovereignty' over its arctic territories. One such was prompted by the tanker *Manhattan*'s transit of the North-West Passage in 1969 – and with good reason, since the purpose of this experimental voyage was to determine whether the Canadian archipelago could become an international seaway for the transport of oil, with the attendant risk of accident and pollution.

Another burst of patriotic protest occurred in August 1985, when the US Coast Guard icebreaker *Polar Sea* informed the Canadian authorities of her intention to make the passage but – according to Canadian accounts – did not formally ask their permission. If this was a deliberate legal challenge, it was swiftly and firmly answered – within weeks Ottawa had announced new legislation to enclose the archipelago as 'internal waters'.

Ottawa's bribe

The history of icebreaking in Canada goes back to 1873, when the residents of Prince Edward Island in the Gulf of St. Lawrence said they would join the Confederation only if the federal government provided a year-round ferry to the mainland. They got their ferry, a wooden steamer, but it was not up to the job, so they appealed right to the top – in a petition to Queen Victoria – for a replacement. A stronger vessel duly arrived from Scotland.

Over the next century, the icebreaking fleet expanded sporadically as new requirements came along: the opening of the Hudson Bay port of Churchill for grain shipments in the 1930s; the construction, with the Americans, of the DEW Line arctic radar chain in the 1950s and the opening of the St. Lawrence Seaway in 1959. The first 'heavy arctic icebreaker' capable of operating anywhere in the Canadian archipelago for at least part of the year, the

13,000 ton *Louis S. St-Laurent*, was built in 1969, coinciding with the *Manhattan's* experimental voyage to the newly discovered Alaskan oilfield at Prudhoe Bay.

The alarm that was caused prompted an urgent, though nevertheless long-winded political debate as to whether Canada also needed a more powerful Polar Class 8 icebreaker, or even a full-blooded Polar 10 vessel, to help control the giant tankers that might soon be pounding through the arctic wilderness. The bigger design would almost certainly have required nuclear propulsion to give her the necessary combination of power and endurance. Canada, in other words, might just have joined the USSR in developing this esoteric technology. But it was not a practical proposition. The Canadian CANDU power reactor, though an eminently sensible design, was not suitable, and the only Western nation prepared to help was France. Besides, who was going to pay for it?

Polar attitudes

In the event, the alarm subsided, along with the oil crisis of the early 1970s, only to be revived by the Americans' seemingly provocative *Polar Sea* transit of 1985. The Polar 8 project was promptly confirmed – but still not funded. The eventual compromise was major modernisation of the *Louis S. St-Laurent*, including a new propulsion unit and an extra section welded into her hull. Yet another refit has since extended her operational life to 2017 – by which time the long-awaited polar-class vessel should be ready to take over.

Forty years of argument was resolved by Prime Minister Stephen Harper on August 28, 2008. Speaking at the Inuit settlement of Inuvik (so far north, as he pointed out, that you look south to witness the Northern Lights), he formally announced this $700 million 'sovereignty' project. The new ship should be able continuously to break ice up to 2.5 metres (8 ft) thick. It will be named after the former Canadian prime minister John G. Diefenbaker – another awkward mouthful for the men on the bridge, but apparently in keeping with Coast Guard tradition.

The choice of name both recognised Diefenbaker's personal record in fostering arctic communities and signalled the current administration's readiness to prepare for an active role in the region's development. Harper coupled his announcement with three others, extending Canadian research, jurisdiction and control in the North-West Passage. 'The ghosts of Hudson, Franklin, Amundsen, Larson, Bernier and the rest', he promised, 'will watch as the *Diefenbaker* crashes through the pack ice'.

NORTH-WEST PASSAGE

Canada's sovereignty in the Arctic is indivisible. It embraces land, sea, and ice.
                    Joe Clark, Secretary of State for External Affairs,
                                                September 10, 1985

In the early days of creative cartography, when map makers included islands and waterways which simply 'must' be there, regardless of the physical evidence, the North-West Passage was much further south than it is today. The mythical 'Strait of Anian' led directly from the Pacific coast of North America to somewhere near Hudson Bay or the St. Lawrence.

The reality is altogether less convenient. As explorers approaching from the Pacific side probed steadily northwards to Alaska, they found nothing until the Bering Strait led through into the ice-strewn waters of the Beaufort Sea. Sailing from the Atlantic side ('North-West' is after all a eurocentric designation), the approach was from the tip of Greenland through Baffin Bay or Hudson Strait. In either case, the crux of the passage crossed what is now Canada's arctic archipelago – 700 nautical miles by even the shortest route.

Counting from a small-scale, broad-brush atlas, the archipelago seems to consist of a dozen large islands, but it actually contains hundreds of smaller ones, some of them nonetheless significant for their location or their history. Most are within the Arctic Circle – that

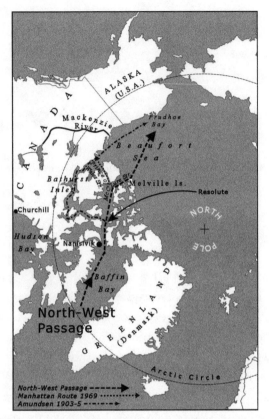

Map 8. Possible routes through the North-West Passage.

is beyond latitude 66 degrees, 33 minutes N – although Canadian arctic maritime regulations start at 60 degrees N, including Hudson Bay – where there is plenty of winter ice.

Inuit know-how

With all the current talk of the Arctic melting, it is easily forgotten that the archipelago is usually – not just occasionally – blocked by ice; so much so that in winter some of the smaller channels can barely be distinguished from the land. The slowly migrating Inuit who lived there for thousands of years before Europeans arrived treated it as a continuous environment, moving seamlessly from snow to ice floe to open water. They survived only by developing an astonishingly efficient lightweight technology – fur clothing, dog sledges and slender kayaks – fashioned from the bones, skins and sinews of the animals they hunted, plus whatever scraps of metal or wood they could find. When the British navy turned up, the Inuit could barely comprehend the luxury of living inside a floating island of massive iron-bound timber (no wonder they were eager to barter or purloin what they could). Yet in the end it was Franklin's men who died, lugging their heavy, unsuitable equipment.

Even today, with sea ice on the retreat, for the North-West Passage to be briefly open in August–September is a notable event, to be celebrated or bemoaned according to one's perspective. Complete transits in any given year are counted in single figures. And news of the route being 'open' has to be carefully qualified, because since there are many islands, there are in theory many passages. Almost any of them may on occasion be used by small vessels supplying remote outposts, or perhaps conducting geophysical surveys. In general navigational terms, however, and hence for the ice reports, just two routes are of major importance.

The winding, relatively shallow passage jointly discovered by Franklin and his rescuers, and eventually navigated by Amundsen, follows the mainland coast. It can be approached from the east either through Fury and Hecla Strait at the far end of Hudson Bay, or from Baffin Bay through Lancaster Sound. It takes in various settlements, crosses the entrance to three rivers and was briefly free of ice in the three years 2007–9.

But the deep-water route that will interest international shipping if it regularly becomes clear – the 'Parry Channel' – runs directly westwards from Lancaster Sound, across Melville Sound and through the McClure Strait, where an alternative branch runs south-west through Prince of Wales Strait. This was the main

North-West Passage referred to with such excitement in 2007, although its opening – not since then repeated – did not coincide with an ice-free North-East Passage. It was also this direct crossing that was attempted in 1969 by the American icebreaking tanker *Manhattan*, on a pioneering voyage that was to have dramatic impact on the Canadian government's policy towards the Arctic.

An icebreaking leviathan

The *Manhattan* experiment addressed three questions that are still pertinent. Could the North-West Passage physically become a practicable seaway for large cargo vessels, not just the setting for an adventurous cruise or a secret nuclear submarine transit? Would it be profitable for the petroleum industry in particular to make extensive use of it? And if so, how exactly would its navigation be controlled?

Reporting on the voyage at the time, the answers already seemed pretty clear. But that was before the accelerating effects of climatic change became evident. Forty years on, in a new physical environment, uncertainties persist. The comparative costs of arctic oil production have to be recalculated, the choice of tanker or pipeline reassessed and the ultimate legal regime determined.

By coincidence, I happened to be aboard the American leviathan some time before her attempted passage, to inspect her experimental 'flume tank' stabilisation, and I remember the captain saying that twin screws and rudders made her surprisingly manoeuvrable – a useful quality in an icebreaker. But that was a trivial factor in her drastic 40-million-dollar makeover from simple oil carrier to icebreaking pioneer.

At 100,000 tons deadweight, the *Manhattan* was the largest merchant ship flying the American flag, so she already possessed something more fundamentally useful in breaking ice – enormous weight. To make it count, however, she needed a sloping bow that would ride up over the ice and push it aside. Exceptionally powerful steam turbines delivering 43,000 shaft horsepower were considered adequate, but the hull had to be strengthened and her vulnerable rudders and propellers as far as possible protected.

Arctic surgery

Shipbuilders take such things in their stride. In this case, they cut the 1,000-foot hull into three vast chunks, shaped a new bow sloping backwards at an acute angle of 15 degrees, doubled up the plating along the waterline, modified the stern and then welded all the pieces back together. She came out 65 feet longer and a useful 9,000 tons heavier.

Late in August 1969 the new *Manhattan* set out from Pennsylvania, bound for the newly discovered Prudhoe Bay oilfield in Alaska. The venture's political sensitivity was already apparent in official exchanges between Ottawa and Washington over matters such as the status of the icebreaker escort – to be provided primarily by Canada's John A. McDonald, supported by a US Coast Guard vessel.

The tanker met her first ice on September 2, brushing aside slabs more than 4 metres (13 ft) thick without effort. All went well until she encountered piled-up, hard-packed ridges in the McClure Strait, where for the past week the wind had been blowing hard from the north-west. At one point, she ground to halt against an 8-metre ridge, and the ice immediately closed behind her. The escorting icebreakers had to move in to extricate her.

The tanker's captain now turned instead into Prince of Wales Strait, running south-west to pick up the coastal passage through the Beaufort Sea, which is often ice-free for a while in summer. Finally, reaching Prudhoe Bay on September 14, she loaded one symbolic barrel of oil before retracing her route through the islands.

Pyrrhic victory

For the *Manhattan*'s owners, the Humble oil company, the voyage had been a triumph, but of a much qualified kind. The most direct North-West passage had proved impassable, even for this monstrous vessel. Without icebreaker assistance, the voyage might have ended disastrously – and unnecessarily, since the alternative route avoiding the McClure Strait was always available. And there were smaller snags. The primitive satellite navigation system available at that time did not always work (at one stage, the crew resorted to the ancient method of measuring speed through ice by throwing blocks of wood overboard). The artificial ice 'islands' against which she might have moored in the Beaufort Sea broke up.

Above all, these difficulties had been encountered when ice conditions were supposed to be at their best. Bringing Alaskan oil out by sea might look profitable on paper – saving an estimated 40–60 cents a barrel – but the view from the tanker's bridge was not so encouraging. The *Manhattan* had to return to arctic waters the next April for further trials in more extensive ice.

These proved disappointing. The tanker sustained some damage to her stern while manoeuvring off Baffin Island and did not attempt another transit. Humble finally admitted defeat and threw in its lot with other oil companies planning a pipeline across Alaska to allow the oil to be loaded from a conventional terminal on its southern coast – the solution still in use today.

The experiment's negative conclusion, however, did not prevent a sudden upsurge of alarm and indignation from Canadians at the prospect of their ecologically vulnerable archipelago becoming a tanker highway. For all the delicate protocol surrounding the *Manhattan*'s passage, it highlighted a fundamental dispute between the two arctic neighbours that is not fully resolved even today. Is the North-West Passage legally an international strait, as Washington would have it, or part of Canada's internal waters, as Ottawa firmly maintains?

Canadian anxiety was only too understandable. In the excitement of the Alaskan oil strike, Humble was not the only company talking about using tankers to ship it out. Others were hurriedly making similar plans, suggesting, for example, that the Canadians might like to turn Herschel Island, an old whalers' refuge just west of the Mackenzie River, into an international oil terminal. There were even mad schemes, seemingly proposed in all seriousness, to build a fleet of submarine tankers. (I cannot see many tanker captains applying for a transfer!) This was 20 years before the *Exxon Valdez* disaster demonstrated what damage a load of spilt Alaskan oil could do, albeit not on the arctic shore, but the Canadian authorities were already alive to the danger – for example, from the grounding in 1970 of a Liberian tanker on the coast of Nova Scotia.

'Inflamed nationalists'

The initial American response was one of unaccustomed bewilderment. In a US State Department briefing for Henry Kissinger just before the second *Manhattan* voyage, an official described it providing tremendous ammunition for 'inflamed nationalists' pressing for a Canadian declaration of arctic sovereignty. Whatever happened now, he argued – a second voyage with or without Ottawa's permission, or just a cancellation – Canadian diplomats would turn it to their advantage. And so they did!

The year 1970 marks the beginning of a cautious incremental campaign, spread over several decades and taking in lengthy negotiation of the 1982 UN Law of the Sea Treaty, to establish Canadian control over its northern waters. The *Manhattan*'s dramatic voyage, and the US icebreaker *Polar Sea*'s contentious transit in 1985, did indeed provide ammunition – prompting the indignant newspaper editorials and public opinion polls to which prime ministers respond. But it was nonetheless a careful legal process.

The first step was an easy one – extending Canada's territorial waters from 3 to 12 miles in line with many other countries. Since the passage taken by the American tanker was in two places less than 24 miles wide, this meant that any vessel following her route would automatically be subject to certain restrictions under international law – as later codified by the UN treaty's provisions for a 'right of innocent passage' through territorial waters.

The law delineating Canadian territorial waters was further amended to allow the Ottawa government to establish exclusive coastal fishing zones. Then in a separate piece of legislation – the 1970 Arctic Waters Pollution Prevention Act – shipping access to the archipelago was controlled by dividing it into separate zones (rather as the Russians do), each requiring a different ice capability, depending on the season.

The 'arctic exception'

These moves were diplomatically followed up through the 1970s by negotiation of an 'arctic exception' from some of the wider provisions of the UN Law of the Sea. Article 234 of the 1982 convention allowed coastal states to enforce non-discriminatory regulations 'for the prevention, reduction and control of marine pollution from vessels in ice-covered areas....where particularly severe climatic conditions and the presence of ice covering such areas for most of the year create obstructions or exceptional hazards to navigation, and pollution of the marine environment cause major harm or irreversible disturbance of the ecological balance' – in other words, in the Canadian arctic.

Ottawa might have had more trouble obtaining this exception had the two most powerful arctic powers, the USA and the USSR, not been hampered by conflicting priorities. The US administration fundamentally objected to anything which restricted American vessels' freedom of navigation – not just through this passage but at any sensitive bottleneck like Gibraltar, the Bosphorus, Hormuz or the Malacca Strait. At the same time, Canada was a close neighbour, trading partner and staunch NATO ally with whom Washington was loathe to pick an argument.

For their part, the Russians fully shared the Canadian government's impulse to exert control over its coastal waters. At that time, they certainly claimed full historic sovereignty in their own arctic waters, and jealously guarded their maritime privacy. Yet they too sought freedom of manoeuvre wherever possible for their rapidly expanding naval forces and merchant shipping.

Canada's boldest move came in 1985, following the politically convenient provocation of the US Coast Guard icebreaker *Polar Sea*'s transit. On September 10, Secretary of State for External Affairs Joe Clark declared that the arctic islands were to be enclosed as historical 'internal waters' with almost immediate effect – in other words, without waiting for friendly US diplomacy to undermine the government's resolve.

Under UN maritime law, this means drawing straight 'baselines' across all the entrance channels so as to turn the archipelago into a single geographical entity, with a continuous boundary. Outside lie the 200-mile economic exclusion zones for oil exploration and fishing. Everywhere inside the baselines is by definition internal waters – over which the coastal state has total control. Foreign ships do not even have the restricted right of 'innocent passage' provided for within territorial waters, let alone the freedom of 'transit passage' through an international strait, where even naval vessels can come and go as they please.

Joe Clark acknowledged that for Canadians this was an emotional issue of fundamental sovereignty:

> Canada is an arctic nation. The international community has long recognised that the arctic mainland and islands are a part of Canada like any other, but the Arctic is not only a part of Canada, it

is part of Canadian greatness ... Full sovereignty is vital to Canada's security. It is vital to the Inuit people. And it is vital to Canada's identity.

Clark rounded off his statement by announcing the construction of a $500 million polar icebreaker – a project which did not in fact materialise for many years, by which time it was estimated to cost $700 million. He emphasised that Canada was not trying to exclude foreign ships by enclosing the islands in this way, merely to supervise their activities:

> Our goal is to make the North-West Passage a reality for Canadian and foreign shipping as a Canadian waterway.

Anticipating that this dramatic, unilateral assertion of sovereignty might not go down too well in Washington, the secretary of state also announced immediate talks on arctic co-operation with Canada's powerful southern neighbour. But this was not enough to placate the Americans. They refused to accept the Canadian legal position, and they were joined in that protest by Britain and other European states.

Such a fundamental dispute cannot easily be resolved, though it may eventually be submerged in the practicalities of jointly managing the North-West Passage. American high concern for strategic freedom of manoeuvre and energy security is not matched in Canada, where protecting the Arctic's fragile physical and social environment is given top priority under the emotive banner of 'sovereignty'. (The statement by Joe Clark from which I have just quoted mentioned it 19 times!)

'Use it or lose it'

Twenty years on, this word continues to have political resonance, but the emphasis has switched from establishing a legal basis for Canada's arctic sovereignty to providing the means to exercise it.

In a typical reference, Prime Minister Stephen Harper said his government understood that 'the first principle of arctic sovereignty is "use it or lose it"'' – by which he meant not just building a new polar icebreaker but the establishment of an arctic military training centre in the Parry Channel at Resolute Bay, a fleet of arctic patrol ships, and the construction of a deep-water naval base at the northern end of Baffin Island, near Lancaster Sound, at the old mining settlement of Nanisivik. Construction should begin in 2011 for completion in 2015. Lancaster Sound has meanwhile been designated a 'marine protected area' for birds and sea mammals (whales, incidentally, do not migrate through this passage).

However, such initiatives are a response to arctic maritime development, not a determinant. Setting aside the handful of vessels which have reason to make a complete transit of the North-West Passage each year – most of them icebreakers – internal traffic serves remote, mainly Inuit, communities, plus the modern bases from which gas, oil and minerals are increasingly exploited. Each summer a variety of coasters, barges and technical support vessels fan out from Atlantic ports, Hudson Bay and the Mackenzie delta.

A political voice

The indigenous arctic peoples' traditional way of life, based on hunting and fishing, has long since been compromised, for good or ill. But they do now have a political voice – in the Arctic Council, in the US state of Alaska, in Canada's arctic territory of Nunavut ('Our Land') and especially in Greenland, where the 2008 referendum produced an overwhelming vote in favour of self-government and possible future secession from Denmark. And this voice often chimes in with a much wider community's concern to preserve a fragile environment, while not being averse to the obvious benefits of economic development.

The minimum pace of that development – and hence the maritime traffic supporting it – will often therefore be governed by local politics. The maximum pace will be a function of worldwide markets for hydrocarbons and minerals, and beyond that the uncertain effects of global warming.

Over the past decade, for example, the pros and cons of building a deep-water port on Nunavut's Bathurst Inlet have repeatedly been debated, referred and delayed. The inlet sits deep along the southern 'Amundsen' branch of the North-West Passage. The idea is to connect the port by an all-weather road to the interior, where a number of companies already mine diamonds, silver, copper and zinc – a case of fuel and other supplies in, minerals out, with new jobs created along the way. But what with changing commodity prices, fears for the calving grounds of the Bathurst caribou herd and the usual funding problems, the project is currently on hold.

There is little doubt, however, about the expanding future for arctic gas and oil; and if melting ice permits, maritime transport along the North-West Passage will play an important role. Even amidst persistent sea ice, the oil men's ingenious technology enables them to drill wells offshore in the Beaufort Sea, from 'ice islands' or specially designed drilling barges. This generates work for survey ships, support vessels, tugs, icebreakers and so on, though not in the first instance for tankers.

When Alaskan oil began to flow from the vast Prudhoe Bay field in the 1970s, a strategic decision was taken – following the dramatic *Manhattan* experiment – to pump it overland by pipeline for loading aboard ships at a more southerly terminal. That decision has not been successfully challenged since then. But pipelines also have their problems, such as the serious corrosion that showed up in Alaska in 2006, and if the ice really does clear for much of the year, the more flexible option of tanker transport will surely be reconsidered.

Mackenzie's oil

In northern Canada, the oil story goes right back to the explorer Alexander Mackenzie, who in 1789 actually saw it seeping from the bank of his newly discovered river. The first wells were drilled right there in 1920, and the oil refined for local consumption or later sent further afield by pipeline.

After the Second World War, the rigs moved further north among the arctic islands, starting in 1961 on Melville Island, near Parry's old Winter Harbour. During the 1970s and 80s, the search extended around the Mackenzie delta and out into the Beaufort Sea (where the important dividing boundary between the American and Canadian continental shelves is still in dispute).

By this time – from 1977 – oil was pumping south from Prudhoe Bay. Canada could not match that, but on its side of the border drilling rigs tapped into large reservoirs of gas. Current plans for extracting it – caribou and grizzly bears permitting – envisage a Beaufort Mackenzie Valley Pipeline running south to connect with the existing distribution network in Alberta (which will generate some associated maritime traffic). But there may later be scope for icebreaking gas tankers, particularly bearing in mind that gas does not pose the same pollution risks as crude oil.

In the late 1970s, considerable research and much speculative money went into a scheme to export Canadian gas by this method. The $1.5 billion 'Arctic Pilot Project' proposed piping gas across Melville Island to a liquefaction plant on the south shore, where it would be loaded into icebreaking refrigerated LNG (Liquid Natural Gas) tankers for shipment eastwards through the Parry Channel, and down the east coast of Greenland to the Atlantic. A pair of these immensely powerful ships, capable of breaking more than 2 metres (about 7 ft) of ice, would keep the export shuttle moving even through the winter.

The project might have worked technically, in which case – especially with the ambitious Dome Petroleum involved – it could have led to similar schemes for crude oil shipment. But in the event, it fell victim to oscillating gas prices and years of political wrangling. Inuit representatives feared that an eventual procession of mammoth ships thudding through the ice would disrupt their hunting grounds, especially by frightening off whales that rely on sound to navigate.

What the whales will hear

As melting ice frees up the eastern end of the North-West Passage, an increasingly self-confident Greenland nation – soon perhaps completely independent of Denmark – is also poised to begin the cycle of oil exploration, extraction and export. Other sources of wealth still hidden in the ground, like the Black Angel zinc mine on the west coast, will surely emerge as the glaciers retreat. Greenlanders, under their geometric red and white flag, will then have to confront for themselves the social and environmental dilemmas of economic development. Tankers will not just thud past, but stop to load. The whales will be listening.

For the moment, although dozens of cruise ships visit Greenland waters each season, traffic through the Canadian islands at the centre of the North-West Passage is extremely modest. A 'snapshot' taken for the Arctic Council's *Arctic Marine Shipping Assessment* in 2004 showed just five complete transits and 107 voyages to destinations within the archipelago. During the exceptionally ice-free season of 2007, when the passage opened for five weeks, there were 12 transits. So the Parry Channel is still a long way from being the mighty international waterway explorers dreamed of, and cautious members of the Canadian maritime community believe it could be decades before it becomes a commercially viable alternative to the Panama Canal, even assuming the melting trend continues.

The problem is that as multiyear ice packed hard against the northern coasts of the archipelago melts, it breaks up and drifts down channels open to the north-west – including the McClure Strait that stopped Parry, and opposite to the southward, the McClintock Channel whose wind-driven ice trapped Franklin. Ice accumulated awkwardly in 2008, for example, delaying the summer clearance in 2009. Given this sort of movement, navigational conditions will be highly unpredictable. So notwithstanding the warming trend, things could actually get worse before they get better.

Arctic bridge

In the meantime, the Hudson Bay port of Churchill is looking way beyond the North-West Passage to build an 'Arctic Bridge' to Russia's Murmansk, at the entrance to the North-East Passage – not just linking up the two trans-arctic seaways, but also providing a possible new trade route between North America and Eurasia. Churchill forms the western end of the bridge, describing itself as 'Canada's only arctic seaport'. Though not within the Arctic Circle, there can be no doubt about its status as an arctic seaport; the approaches are choked with ice for much of the year.

Harbour and railway were built by the government in the 1930s to export grain and that, in the main, is what Churchill still does. But it also offers deep-water berths for tankers and general cargo vessels, promoting itself as a commercially attractive short cut between the landlocked midwest of the USA and northern Europe, saving shippers time and money by cutting out the St. Lawrence Seaway. Through ice-free Murmansk, at the other end of the 'bridge', it could potentially connect with railways, rivers and canals leading right across Russia to the heart of Asia.

Domestically, Churchill is an important supply base for arctic mining operations, just as it was for Cold War military installations. During the Second World War, it even sent supply ships as far as the Soviet Union, but until November 17, 2007, when the Murmansk Shipping Company's *Kapitan Sviridov* docked with a cargo of fertiliser, it had never received an inbound Russian shipment.

The fertiliser had been loaded in Estonia, and was on its way to a farming co-operative in Saskatchewan. Two more Russian shipments arrived the next year but nothing in 2009 – although a Russian cruise ship did call.

Not just polar bears

Churchill's normal navigable season, during which ice-strengthened ships are not required, lasts from mid-July to early November. The expectation is that a warming climate will gradually extend this by several months, enabling the port – now owned by OmniTRAX – to diversify and improve the balance of its traffic. Government support to upgrade port and railway facilities was announced shortly before the *Kapitan Sviridov*'s arrival.

Three ships do not make much of a bridge, but the Russians are certainly keen on the idea. The Murmansk Shipping Company's director general Alexander Medvedev said that once the new route had been proven, and without waiting for climatic change, they could consider extending the shipping season 'with ice-strengthened ships using lanes opened by icebreakers'. In response, the port development authority politely explained that it could always call on the Canadian Coast Guard if icebreaking was required; a Russian offer would be 'appreciated', but would involve waiving current Canadian legislation governing foreign vessels.

For a small community in the Canadian wilderness – population 1,000 – the Arctic Bridge is an ambitious, imaginative vision. What becomes of it will be determined by cold calculations of comparative transport costs and the still uncertain speed of global warming. Until then Churchill will have to make the best of its solid seasonal trade in exported grain and its recent new venture into eco-tourism – 'polar bear tours'.

NORTH-EAST PASSAGE

Russia's power will grow with Siberia and the Northern Ocean.
Mikhail Vasilyevich Lomonosov,
eighteenth-century Russian scientist and poet

Shortly before *glasnost* began to expose the fatal weakness of the Communist empire, collapsing like a glacier afflicted by global warming, the Soviet leader Mikhail Gorbachev travelled north from Moscow to expound a dramatically new vision of the Arctic Ocean's development. And the setting he chose, appropriately enough, was Murmansk.

On October 1, 1987, in one of those all-encompassing speeches in which he liked to indulge, Gorbachev called on all the USSR's maritime neighbours to make the Arctic 'a zone of peace'. He acknowledged that since most of those neighbours were members of NATO, at that time still in residual military confrontation with the Warsaw Pact, this was an ambitious aspiration. But the Soviet Union, he said, was seeking 'a radical reduction in military confrontation' and would welcome discussion of the region's long-standing security issues.

Meanwhile the Soviet leader made a number of sweeping proposals to reduce tension and foster co-operation: transformation of northern Europe into a nuclear-free zone; reduction of naval operations in the Baltic, North, Norwegian and Greenland seas; co-operative exploitation of the Arctic's resources, such as oil and gas; a conference

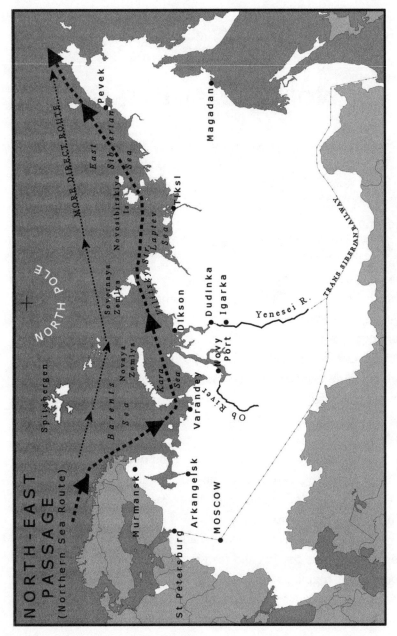

Map 9. Routeing options through the Nort-East Passage.

to co-ordinate scientific research; and a joint plan to prevent environmental pollution.

The announcement that caught seafarers' attention, however, was a tentative offer to reopen the 'Northern Sea Route' (NSR) to foreign ships, with the assistance of Soviet icebreakers. When a similar invitation was issued 20 years previously, there had been little positive response from the international shipping community. The Japanese were alone in showing serious interest, only to be deterred by the opaque Communist bureaucracy they encountered. But this time the prospects looked much better. Gorbachev was now promising transparency, the whole political atmosphere was more relaxed, domestic shipping traffic along the Siberian coast was at a peak and the nuclear-powered *Sibir* had just demonstrated the Soviet icebreaking fleet's formidable capability by conducting a complex research programme at the North Pole in May and June, when the arctic sun barely had time to weaken the ice.

In the event, the route was not formally opened to foreign ships until July 1, 1991, by which time the USSR was beginning to disintegrate and with it the elaborate infrastructure on which the route's functioning depended. In short, this second official opening was another anticlimax. But it did eventually set in train serious international study of the many technical, legal and commercial problems involved in opening up this arctic seaway, so that when global warming suddenly seemed to make it a physical possibility, much of the preparatory work had already been done.

Baffling statistics

To understand the often baffling Russian statistics, one needs to distinguish between the different terms used to describe their arctic waters. Historically, explorers dreamed of a 'North-East Passage' from Europe to China and the Indies – and the Russians also thought of it in that way, on a global scale, although in the first instance they were eager to discover and exploit the resources of their own Siberian land mass. Today's Russians talk of 'Sevmorput' ('The Northern Sea Route'), which for them is both more and less than the North-East Passage.

The Northern Sea Route's present administrative definition excludes the busiest section of the passage, the Barents Sea. It only starts at the entrance to the Kara Sea, and stretches eastward through the Laptev and East Siberian seas to the Bering Strait. Confusingly therefore, much of the shipping traffic actually plying the Arctic does not appear in official NSR statistics. The authorities have proposed extending the definition, and some of the relevant jurisdiction, westwards as far as Murmansk, but this has so far been resisted by oil and gas interests in the region. For the time being, when Russians are already referring to the whole passage, including the Barents Sea and beyond, they tend to use some broader term like 'maritime transport corridor'.

In another sense, however, the term 'Northern Sea Route' embraces more than the North-East Passage because it links up with a vast river network penetrating the Siberian taiga east of the Urals, most notably along the Ob, the Yenisei, the Lena and their innumerable tributaries. And since railways and all-weather roads are scarce in these remote areas – expensive to both build and maintain, especially in a warming climate – river transport is enormously important.

From the north, seagoing ships can and do penetrate deep inland. For example, Novy Port and Igarka, respectively, on the Ob and the Yenisei, are something like 400 miles from the open sea. In British terms, it's as if cargoes bound for Edinburgh or Glasgow came in through the Thames Estuary.

But far beyond these estuaries, smaller waterways stretch southwards, eventually linking up in a few places with the Trans-Siberian Railway. Fleets of small shallow-draft vessels and barges shuttle back and forth with construction materials, fuel, products for export and supplies of all kinds. Some river craft also take to the sea for short hops along the arctic coast, competing for custom with seagoing vessels, or load and unload from large ships anchored in the estuaries.

None of this is easy work. Siberian rivers are frozen solid for much of the year. When the spring thaw comes, they are dangerously in spate, after which some of them almost run dry. With all these problems, the navigable season may last only a few months and in some places is measured in days.

The NSR definition which makes most sense in modern economic terms is therefore the broadest – a transport corridor running from the Atlantic to the Pacific, opening up Russia's northern industrial centres – and especially the oil and gas fields around the Barents Sea – to Western markets and foreign investment. The limited official definition harks back to the early Stalinist period, when Glavsevmorput was created not just to operate ships, but to administer the region's forcible industrialisation – literally so in the sense that forced labour was so often used. That bloated Communist organisation was disbanded in 1964. Traces of the old bureaucracy may admittedly survive (the re-establishment of a new independent agency to manage the route was announced in 2009) but in a new form and devoted to quite different ends.

Ice cellar

Compared with a pilot book covering, say, the English Channel, navigational descriptions of the Northern Sea Route make for an alarming read. 'The shipping conditions in Kara Sea are extremely hard,' says an introduction to this notorious 'ice cellar' produced by the Russians for an EU-funded research project. 'The main reasons which make the shipping there so difficult are: large quantity of underwater threats, frequent fogs, nearly constant presence of ice and little information about the streams.' Not much encouragement there!

The northern route crosses a long continental shelf, littered with islands. Many of the coastal straits and passages are shallow, limiting the size of ships to about 50,000 tons deadweight. Large areas have not been charted to modern standards – inevitably so, since the route stretches for thousands of miles and is inaccessible for much of the year – although the cash-strapped Russian authorities are doing their best to improve the situation. Navigational aids – lights and beacons – are often inadequate, putting a high premium on global satellite positioning (GPS).

Poor visibility is a frequent problem, particularly where low-lying snow-covered shores do not register on radar, or a solid island can be mistaken for an icefield. A navigational note about the East Siberian Sea warns that on one in three days during the summer season, there is fog, rain or snow. And of course the ice is everywhere.

The sea freezes from October through to May (though across the ocean as a whole, satellites show the maximum ice extent in March), then melts for the next four months or so to leave a maximum amount of clear water in September. Even when a channel opens, it may be obstructed by fields of ice so permanent that they actually have names, rather like mobile sandbanks – the Taimyr and Yansk packs in the Laptev Sea, for example. There are eight or nine of these recognisable 'massifs' along the Siberian coast. The exact course a ship plots is likely to be determined as much by the ice distribution as by the depth of water or the availability of navigational marks.

Late developer

The crux of an NSR passage, roughly half way along, is the Vilkitsky Strait, which takes vessels from the Kara to the Laptev Sea past the northernmost cape of the Siberian land mass. It is at this point that even in a good year, when the ice has retreated a long way, an obstinate tongue tends to cling to the islands of Severnaya Zemlya (not discovered, astonishingly, until 1913). This is where ships are most likely to need icebreaker assistance.

As with the North-West Passage and its multiple channels, the Northern Sea Route is actually a complex of possible routes, depending on which end a ship starts from and what she carries – pipes for an oilfield, fuel for some remote settlement, metals from Dudinka or coal from Pevek. But given clear water (or an icebreaker), the navigator has three main options.

Although Siberian coastal waters do not contain a dense maze of islands like the Canadian archipelago, the shipping route is interrupted by three groups of islands – Novaya Zemlya ('New Land'), Severnaya Zemlya ('Northern Land') and the Novosibirskiye Ostrova ('New Siberian Islands') – separating four distinct seas: Barents, Kara, Laptev and East Siberian. Ships can either stay inshore, moving from one sea to another through narrow straits like the Kara Gates between Novaya Zemlya and the mainland, or go outside some of the islands. Ice permitting, a foreign vessel using the NSR to cross from the Atlantic to the Pacific could head almost directly from the North Cape of Norway to the Bering Strait, outside all the islands except Franz Josef Land.

No need to queue

This offshore option may soon be important because Russia, claiming – like Canada – to base its legal position on the 1982 UN convention, has enclosed the coastal islands by straight baselines. The straits providing the inshore route are therefore defined as historic 'internal waters', completely under Russian control, with no automatic right of 'innocent passage' for foreign ships. The USA and other maritime nations may not like this, any more than they like the Canadian position, but it is difficult to see what they can do about it. And for the moment, the argument is in any case largely academic. Foreign container ships and aircraft carriers are not exactly queuing up to cross the Arctic Ocean.

When the former Soviet government sought to extend its jurisdiction in the Black Sea, US Navy warships were quick to 'show the flag' and assert their right of innocent passage. But that dispute concerned territorial, not internal, waters. Along the Northern Sea Route, as in the North-West passage, the issue is whether the Russian waterway should technically be regarded as a series of 'straits used for international navigation', overriding local jurisdiction – after all, Moscow has repeatedly invited foreign shipping to use the route as a short cut to the Far East. If and when the North-East Passage becomes a substantial reality, this is something the lawyers will have to sort out, though by that time legal principle will probably have been overtaken by commercial practice.

As Russia's modern icebreaking fleet began to take shape, including the early nuclear ships, the first important date in the post-war development of the NSR was 1978. In that year, the important route between Murmansk and the Yenisei river port of Dudinka, serving the Norilsk mining combine, was established on a year-round basis. The cargo it carries – exported ores, nickel and copper, plus local supplies – now amounts to well over a million tonnes a year.

In the early summer of that same year, an ice-hardened Russian freighter escorted by the nuclear icebreaker *Sibir* lifted cargo from Murmansk to the Far Eastern port of Magadan in the Sea of Okhotsk (formerly a gateway to the *gulag*), to demonstrate the feasibility of a complete arctic transit in heavy ice conditions. Transit traffic built up through the 1980s to reach a modest peak of 30 shiploads totalling 200,000 tonnes in 1993, but then fell away completely by the turn of the century. A sudden hike in icebreaking charges – they more than tripled in 2003 – did nothing to restore the situation. And although the route was formally opened to foreigners in 1991, only a handful of ships – Finnish, Latvian and German – have so far taken up the invitation.

Activity within the arctic seaway also declined sharply as Russia's Communist command economy collapsed, to be replaced only gradually by commercial alternatives. Time-expired ships were not replaced. And just as the effects of global warming are amplified in this region, the remote arctic communities were doubly affected by political change. The subsidies and upgraded salaries that had supported them lapsed; jobs disappeared; towns began to empty (the population of the timber port of Igarka, for example, halved from 19,000 in 1989 to 9,000 in 2002). All this was then reflected in the falling demand for coastal shipping.

So, in retrospect, Gorbachev's Murmansk speech, far from signalling the start of a boom period for the NSR, marks a high point from which the level of activity rapidly receded. That year, 1987, almost seven million tonnes of cargo moved along the route. By the mid-1990s, the amount had fallen to about two million tonnes, and it has remained at roughly this level. The latest available figures show that 2.2 million tonnes was shipped to NSR ports in 2008, including 0.6 million tonnes of oil and oil products, but *not* including the large volume of oil – about 10 million tonnes a year – nowadays moved around the Barents Sea, because that was not yet officially part of the route.

Second refusal

The fact that the international shipping community ignored the offer of a short cut through the Arctic naturally disappointed Russian officials who had hoped to earn badly needed foreign currency from the sale of icebreaking, pilotage and other navigational services. But if they felt snubbed, they managed nonetheless to turn the situation to positive advantage by taking an active part in an unprecedented multinational research programme to discover what had gone wrong. Was the Northern Sea Route technically feasible for foreign ships, and if so, was there real commercial potential in using it, either within Russian waters or as a North-East Passage?

The *International Northern Sea Route Programme (INSROP)* was mainly a Norwegian-Russian-Japanese investigation lasting six years, 1993–9, enlisting experts from 11 other countries including Canada, Finland, Germany, the UK and the USA. Nothing remotely like it had been attempted before. It set out to study four main aspects of the problem:

- physical conditions along the NSR and the problems of ice navigation
- environmental factors
- the potential for trade and commercial shipping
- political, legal and strategic factors

No less than 167 reports were eventually published, and as one would expect, given the purpose of the exercise, a great deal of detailed information the Russians had previously kept to themselves was disclosed.

INSROP concluded that regular, extended use of the Northern Sea Route by international commercial shipping *was* indeed feasible, if only to serve the burgeoning oil and gas fields. Ships would need the assistance of icebreakers in winter conditions, and ice-strengthened hulls would always be desirable even in summer, though not always essential. Full, independent transits of the route were more problematic, but certainly possible – a conclusion backed up in August 1995 by an experimental voyage from Yokohama to Kirkenes in northern Norway, using a 15,000-ton ice-class Russian freighter which took only 21 days, including diversions for various scientific purposes.

The study registered that the arctic environment was particularly vulnerable to pollution, but judged that notwithstanding Russia's past record in this respect, its representatives now seemed to realise that the problem must be taken seriously at every level.

A substantial export trade from the western arctic region was identified (potentially perhaps 30–35 million tons a year including oil, non-ferrous metals and timber) and a smaller potential from the eastern region (a few million tons, including apatite, timber, coal and tin). As for complete transits between the Atlantic and the Pacific, research showed that at least on paper, there was a potential cargo base of 5–6 million tons in an easterly direction and 2–3 million tons going the other way.

Practical doubts

It was when the *INSROP* teams began to look at the practice, rather than the theory, that doubts crept in. Ship owners are understandably cautious about changing their operations – and risking investment running into hundreds of millions – unless there is some clear long-term advantage in doing so. In this case, there would obviously be additional costs – icebreaking fees, ice-strengthening and extra insurance – with many uncertainties.

How reliable would the Russian support services be (for navigational information, weather forecasts and ice reports), especially in such difficult economic circumstances? Would central governmental funds be available to maintain such an expensive icebreaking fleet, and if so, would the fees be pitched prohibitively high so as to recover the total cost? How reliably could ships be scheduled, if at all, when subject to unpredictable ice conditions? Might cargoes be damaged by extremely low temperatures? Would there be arguments about jurisdiction or regulation?

It was the same with the vital question of insurance. The international marine insurance market was evidently ready to provide the necessary cover if requested, but needed more information. And since there was little recorded experience of this unfamiliar environment, risks would initially have to be underwritten on a case-by-case basis.

Among foreign ship owners, only the Finns and Latvians had so far ventured into these difficult waters. Where others were concerned, *INSROP* showed there would certainly be no rush to abandon the Suez Canal in favour of the North-East Passage. Owners of the huge container ships plying the southern route between Europe and the Far East were not ready to risk them in the Arctic, even if it did offer time and distance savings. At the other end of the spectrum, however, there seemed no reason why firms interested in occasional low-value bulk cargoes, on voyages where strict scheduling was not an issue, should not take advantage of the Russian offer.

Turn of the tide

The *INSROP* study was completed in 1999, with shipping activity along the northern routes at a low ebb. It had given the outside world access to the Russians' long maritime experience in the Arctic and shown that transits were technically feasible if not commercially attractive. But this was essentially an academic exercise. When geologists signalled that vast deposits of gas and oil were hidden beneath the Arctic and Russia began gradually to develop a mixed economy more capable of exploiting them, as well as reviving older trades, a more practical study was commissioned, focussed on the Barents Sea.

The *Arctic Operational Platform Project (ARCOP)*, 2003–6, was largely funded by the EU, no doubt with an anxious eye on its own energy security, and co-ordinated by the Helsinki-based Aker Finnyards. This new study did not change the fundamental *INSROP* conclusions but looked more closely at the area where oil companies were already sponsoring shipping development, to assess the port facilities and navigational support that would increasingly be required.

Environmental protection – the prevention or at least containment of oil spills – was given a high priority, and here there was reassurance on one point: new generic construction standards introduced by the International Maritime Organisation (IMO), and strongly endorsed by the EU, would phase out most single-hulled tankers within the next few years – that is, vessels with only one layer of steel plating to prevent their cargo spewing into the sea in the event of collision or grounding.

Where the Arctic was concerned, IMO was encouraged to act – and more directly so the US regulatory authorities – by the disastrous *Exxon Valdez* grounding in 1989. Although this was not strictly an arctic event, and not caused by ice, it spilt millions of gallons of Alaskan crude into Prince William Sound, devastating the local fisheries. As always, the improved operating standards took a while to negotiate, but by 2002, the organisation had approved *Guidelines for Ships Operating in Arctic Ice-covered Waters*, albeit only as a set of recommendations covering such things as construction standards, safety equipment and the training of 'ice navigators'. In parallel with this work, the shipping industry's classification societies (which set construction standards rather as domestic building regulations govern house construction) developed *Unified Requirements for Polar Ships*.

Living on borrowed time

In 2005, the Arctic Council nations (Canada, Denmark, Finland, Iceland, Norway, Sweden, the Russian Federation and the USA) weighed in with a comprehensive four-year study, the *Arctic Marine Shipping Assessment (AMSA)* covering the whole of the polar region. As mentioned earlier, this started by taking a 'snapshot' of shipping activity, then examined some of the industrial, commercial and social pressures driving it, and recommended ways in which it can develop successfully, and above all safely. Oil spills, not surprisingly, were again identified as the most significant environmental threat.

The assessment was chaired by the American icebreaker captain–turned-academic Professor Lawson W. Brigham and was notable for the fact that in addition to the usual expert 'workshops', 14 town hall meetings were held to sound out the views of arctic communities affected by this changing environment. The recommendations of its 2009 report were grouped under three main headings:

- enhancing arctic marine safety
- protecting arctic people (of whom there are about four million, including more than 30 indigenous groups) and the environment
- building the arctic marine infrastructure (although along the coast of Norway and North-West Russia, it was already considered adequate)

The sponsoring governments were urged to support the IMO's work on raising standards and to make its arctic guidelines mandatory. A multinational search and rescue system should be created, *AMSA* agued, and protected areas designated. Charts, port services, marine communications, ice-navigation training and the availability of ice reports were all in urgent need of improvement. One of the report's authors, Ben Ellis, from the Institute of the North, warned that without action in some areas, the maritime community was living on borrowed time – 'I just hope it doesn't take a major cruise ship or tanker disaster to get our attention.'

AMSA's shipping survey in the summer of 2004 showed much more activity on the Russian side of the ocean than along the North-West Passage, and heavily concentrated in the Barents and Kara seas, containing the old established ports of Murmansk and Arkhangelsk and the important rivers Ob and Yenisei. This is also where the new oil and gas traffic will develop.

There were 165 voyages in the Kara Sea involving 52 ships, with hundreds more criss-crossing the Barents (compared with 107 in the Canadian archipelago). A handful of icebreakers and research vessels were registered in the central ice pack, but again, no complete transit of the Northern Sea Route.

Eco-tourism

An important new component was the cruise ships, many of them sailing north from the Caribbean to let passengers admire Greenland's icebergs, scan the bird cliffs with their binoculars and hope to catch sight of a whale or a polar bear. The survey identified 27 passenger vessels in Greenland waters that year, and the numbers have continued to grow each season at least until 2007, when there were about 250 in the Arctic as a whole, mostly off Greenland. These luxurious but deliberately somewhat adventurous ships are no more immune from accident than any other ship, as passengers who had to take to the lifeboats during a recent Antarctic cruise will testify, and they pose a major challenge for those hoping to establish an organised search and rescue capability.

The small vessel which sank in the Antarctic – in calm weather – had just 100 passengers on board. Imagine trying to evacuate thousands of mainly elderly passengers from a vast, modern, slab-sided vessel, listing heavily with a big sea running. Even assuming that could safely be achieved in the time available (lifeboat drills tend to be practised in calm water), in present circumstances survival might well depend on the coincidental presence of other ships. The industry is obviously aware of the problem, but nobody wants to spoil the party by talking about it too loudly. As one senior cruise ship's officer put it – 'They are giving it a thorough ignoring, because sometime it is going to happen!'

Overall, the Northern Sea Route and its proposed extension into the Barents Sea present a picture of booming activity in the West – driven principally by oil and gas – and continued depression in the East. It is not yet a North-East Passage.

Russian seafarers respond to this situation quite differently to those looking in from outside. As in Canada or the USA, of course, there must be many Russians for whom these remote regions are simply out of sight and out of mind. But for those involved – for example, the Central Marine Research and Design Institute (CNIIMF) in St. Petersburg, nowadays a private foundation but with close links to the Ministry of Transport – attitudes are conditioned by history, national pride and unconscious habits developed through generations of centralised Communist bureaucracy. Russia is still in transition. The Bolsheviks' 'five-year plans' may long since have been abandoned, but future aspirations are still often expressed in terms of targets, strategies, programmes and doctrines – for example the *Marine Doctrine of the Russian Federation to 2020*, or the *Transport Strategy of the Russian Federation to 2030* – even if what actually happens is determined by who puts up the money.

Russians are rightly proud of their long and in some ways unique experience of arctic navigation – the Murmansk Shipping Company, for example, was established more than half a century ago, and with its subsidiaries now has a fleet of 300 specialised seagoing ships and river craft. Those operating them have done their best to preserve their professional expertise as 'ice captains', polar hydrographers and so on through recent hard economic times, and are naturally eager to make more use of it.

In doing so, they are constrained by an inherited structure assembled over decades of Soviet rule – including remote settlements along the eastern sections of the Northern Sea Route that are heavily, and in some respects totally, dependent on waterborne supplies, and a unique nuclear-powered icebreaking fleet requiring costly maintenance, shore support and re-equipment (not to mention the problem of nuclear waste disposal). The settlements rely on supply convoys, which in turn rely on extensive icebreaking assistance. Both are the product of central direction rather than independent commercial development and continue to require subsidies that may or may not be forthcoming in the future.

Conquest and assimilation

Long before the Russians arrived, let alone the Soviet commissars, these arctic shores and Siberian forests were occupied by scattered groups of indigenous people – Saami (Lapps), Samoyeds and Chukchi, the equivalent of North American Eskimos and Indians – who survived by reindeer herding, hunting and fishing. Initially, Russians from the south knew so little about them that it was popularly believed they died or hibernated in November each year and revived the next April.

The Czars engaged in crude, often violent, colonisation, extracting taxes in the form of furs. After 1917, there was much wider exploitation of the region's natural resources, often using forced labour provided by criminals and political prisoners, and accompanied, for the indigenous locals, by a systematic programme of political and social assimilation. Schools were opened to teach Russian. Herdsmen were taught the supposed benefits of Soviet collective farming. Modern hunting equipment, vodka and some primitive healthcare were supplied in exchange for exposure to Communist indoctrination. Other arctic nations were engaged in a similar process of assimilation, leading almost unavoidably towards cultural annihilation, but none struck such a harsh bargain (one of the first jobs Lenin gave Stalin was 'Commissar of Nationalities').

After many decades of such human and financial investment, profitable or otherwise, there is natural reluctance on the part of Russian officialdom to dismantle the structure and start again – it is easier to adapt what is there and hope to attract more customers for those icebreakers. Besides, the world's seemingly insatiable demand for energy promises to make anyone who can provide it extremely rich.

Great expectations

In September 2008, as if to confirm that not just Western nations are drawn to the Arctic by – in Putin's derisive words – 'the smell of oil and gas', the Russian Federation's security council met in a remote outpost on its northernmost archipelago, Franz Josef Land. Short of boarding an icebreaker, they could not have got nearer to the North Pole.

The meeting was clearly intended to make a political statement for both domestic and international audiences – as no doubt was the Canadian cabinet meeting at Inuvik (Latitude 68 degrees N) that same summer. It was quickly followed by a keynote Kremlin speech in which President Medvedev declared the Arctic 'a region of strategic importance' which should become Russia's 'resource base for the twenty-first century'.

To achieve this, the President said, Russia needed a solid legal and regulatory framework for its arctic activities. The borders of its continental shelf must be fixed in law (an apparent reference to the contentious UN negotiations) and the developmental gap between North and South reduced.

For the maritime community, the speech contained significant references to the arctic region's transport infrastructure, and in particular the Northern Sea Route – which he said should be developed as a major transport corridor, especially for hydrocarbons. Resolving the Arctic's problems, Medvedev added, would require 'a harmonious combination of the state, the business community and local self-government' – in short, a mixed economy.

On the same occasion, the president approved yet another long-winded statement of official doctrine – *Basics of the Russian Federation Policy in the Arctic to 2020 and Beyond* – later published in *Rossiyskaya Gazeta*. This set out broad planning priorities for social and economic development, military security, environmental protection, scientific research and international co-operation. Transport within the 'arctic zone' over which Russia claims sovereignty received prominent mention.

'A resource base for the twenty-first century'

Medvedev was stating the obvious. Whatever the shape of Russia's economic future, it will depend to a large extent on successful exploitation of its arctic resources, especially the new-found gas and oil. That will require a substantial measure of genuine international collaboration, which in turn demands legal clarity concerning ownership and regulation – foreign oil companies want to know where their taxes will be paid.

Post–Soviet Russia's mixed economy will expect less bureaucracy. State-controlled enterprises like Gazprom operate alongside privately owned companies such as Lukoil, which has an American partner; commercial ship owners have to deal with maritime authorities still administered by government, but will want wherever possible to shake themselves free from central control.

Efficient transport – which in this region so often means marine transport – is imperative for social as well as economic reasons. If, as expected, management of the Northern Sea Route is to remain a state function under a new independent department, that will have to get to grips with a whole raft of demands from the maritime community – more accurate charts, better communications, a well-organised search and rescue capability and convincing plans to control oil spills, not just new icebreakers.

If one is to believe the St. Petersburg marine research institute, all this is in hand, or soon will be, provided Moscow comes up with the funding. Among the promised improvements are:

- updated charts – 200 of them in English – incorporating sonar surveys of the main estuaries
- sailing directions in English and Russian
- electronic charts for use in upgraded satellite navigation systems
- three new polar-orbit radar satellites to monitor the ice pack
- an extended weather forecasting network
- search and rescue bases in Murmansk, Arkhangelsk and Pevek
- better equipment for cleaning up oil spills
- new emergency radio stations
- additional hydrographic survey vessels, rescue tugs and diving ships
- revised navigational regulations – to be extended into the eastern part of the Barents Sea and the western Bering Sea
- more competitive tariffs for icebreaking and pilotage services

New icebreakers have also been promised, notwithstanding the two-year 'icebreaker gap' about which CNIIMF recently warned. Three nuclear-powered vessels are being built to a pioneering variable-draft design – drawing between 8 and 10 metres – so as to cope with both river and open-sea conditions. With this flexibility, they are intended to replace five older ships, including the shallow-draft vessels working on the Yenisei. In addition, five somewhat less powerful diesel-powered icebreakers are planned.

The high cost of maintaining this capability is supposed to be borne by the state, which makes the recent transfer of the nuclear fleet from the Murmansk Shipping Company to the nuclear agency Rosatom doubly significant. Although the move has obviously involved considerable upheaval and departmental realignment, one would expect the state agency to have more leverage than a private shipping company in extracting the promised subsidies from central government.

What with changing definitions and overlapping categories, disentangling the various Russian projections of future shipping traffic is not at all easy. They all indicate rapid growth, however, especially where oil and gas are concerned. Russia's oil production steadily increased from the turn of the century and within the next decade is forecast to amount to more than 500 million tons a year. Much of this is exported through pipelines, but a considerable amount goes by rail to White Sea ports such as Arkhangelsk, where it is either shipped abroad directly, or taken round in small shuttle tankers to a transhipment complex in Kola Bay. From there, giant deep-sea vessels of up to 300,000 tons carry it to European and American markets.

As more arctic oil and gas is extracted, onshore and offshore, Gazprom estimates it may eventually require more than 50 oil tankers and other specialised vessels, plus at least a couple of dozen liquid gas (LNG) carriers – the latter designed to ease the load on the pipeline system and increase commercial flexibility. In 2008, the company's new business strategy indicated that it was already considering an order for eight LNG ships to service operations on the Yamal Peninsula and elsewhere.

Ice shuttle

The use of 'shuttle' tankers is very much a pattern for the future. More of them are needed to service new terminals being built in the ice-strewn eastern corner of the Barents Sea, and the vast Kara Sea estuaries beyond. The River Ob already features a double shuttle – from pipeline into 2,000 ton river tankers, which transfer it to 20,000 tonners for shipment to a huge storage tanker in the Kola Bay. All these ships need to be ice-hardened, but those now building for the offshore Varandey and Prirazlomnoye terminals are almost icebreakers in their own right – albeit supported at the terminals by small specialised icebreakers provided by the oil companies. As with the new double acting freighters serving the Norilsk combine on the Yenisei, the operators are evidently seeking as much independence as possible. And the same will no doubt apply to future gas tankers.

The challenge facing the new Northern Sea Route administration, therefore, is to create a modern transport system capable of handling at least 40 million tonnes of cargo a year by 2020. Perhaps two-thirds of this will be oil and gas, for which the state-owned shipping company Sovcomflot reckons to build about a million tons of tankers and gas carriers. Trade in dry cargoes such as metals, ores, coal and fertilisers is expected to continue developing right across Siberia, and there is talk of resumed timber exports from Igarka on the Yenisei, and Tiksi, further east in the Laptev Sea, as well as from White Sea ports – in which case a new fleet of 15–20 timber carriers may be needed. Then there are the 'socially significant cargoes' the authorities are bound by their own transport strategy to deliver – the food, fuel and other supplies without which remote arctic settlements cannot survive, amounting to about 15 million tonnes a year – for which four new ice-class coasters are promised. And finally, there will be a small number of Russian transit cargoes, totalling perhaps half a million tonnes a year.

In all, about 60 new Russian vessels, subsidised to fly the national ensign, are expected to be launched for these trades over the next decade. Half of these will be tankers of one kind or another, the rest container ships, bulk carriers and general cargo freighters, all ice-strengthened to some degree.

Mother of all icebreakers

To make their operation reliable, especially during the eight months of the year when ice is forming, the polar icebreaking fleet, both nuclear and diesel-powered, will be maintained. And if the St. Petersburg institute's hopes are realised, they will be joined in about ten years time by a giant new nuclear design, twice as powerful as the existing *Artika* class. This mother of all icebreakers should be able to 'operate reliably and efficiently under any ice conditions in any area of the arctic basin, including year-round pilotage of vessels transiting high latitudes of the NSR'. In other words, it could lead suitable convoys straight across the Arctic Ocean, ignoring the coastal passages, even in winter.

The big question is who ultimately will pay for it, or indeed for the other new icebreakers? The nuclear fleet's construction and operation are already subsidised by the state, and its income from icebreaking fees was reduced by Norilsk's decision to build its own fleet of ice-capable freighters. So where is the new business?

One hope to which the Russians cling – in spite of past disappointments – is that with the prospect of the ice melting, foreign ship owners will suddenly realise the advantages of a short cut to and from the Far East, avoiding Suez and the Somali pirates. The domestic Northern Sea Route will expand into an international North-East Passage, although until ice conditions substantially improve, Russian analysts are not expecting much more than about half a million tonnes of transit traffic a year.

Judging from the response to the *INSROP/ARCOP* studies, however, one thing is clear: foreign shipping will not be prepared to pay the full costs of services created to meet the strategic needs of Russia's northern economy. Nor will foreign owners expect to be charged regardless of the assistance they actually require. The sharply increased fees in 2003 are quoted as the main reason why Russian transits came to a halt. Since then the nuclear fleet's management has been transferred from the privatised Murmansk Shipping Company to Rosatom. If the state agency is serious about attracting foreign custom, it must design a flexible tariff and strike a sensible balance between the need to contain subsidies and setting fees at a commercially attractive level.

In summary, Russian strategy for the Northern Sea Route has changed remarkably little over the past 40 years, even though the regime directing it from Moscow has been transformed by the collapse of the Soviet Union and the cargoes moving along it are in many respects quite different. The completely new element is the prospect of an ice-free North-East Passage throughout the summer months. So just how tempting is that likely to be? How far can the dream that inspired the early explorers now be realised?

Across the Top of the World

Perhaps I should apologise for this heading, whose meaning is clear yet at the same time misleading. Cartography has indulged my habitual image of the world – in this case, encouraging the habit of thinking that North is 'up', so that the Arctic Ocean must be on top. Ships taking a North-West or North-East Passage will have to struggle *up* the world, across and *down* the other side – though in doing so, they can save thousands of miles by comparison with heading south-about through the Suez or Panama canals, let alone round one of the southern capes.

A British atlas which, while acknowledging that the world is round, nevertheless spreads out the northern continents across the page – with Europe conveniently in the middle – introduces another misunderstanding. Now the western tip of Alaska and the eastern extremity of Siberia appear to be on opposite sides of the world, whereas in reality they are so close you can see one from the other.

As a yachtsman, I am also conscious that Mercator's clever projection can be misleading, spreading apart lines of longitude towards the Pole so that arctic distances look greater than they really are. Marine charts adjust for this by spreading the parallels of latitude proportionately, so as to preserve the shape of the coast-line. (One of the first things you learn in navigation class is always to measure distance from the side of the chart, at the approximate latitude of your own vessel.)

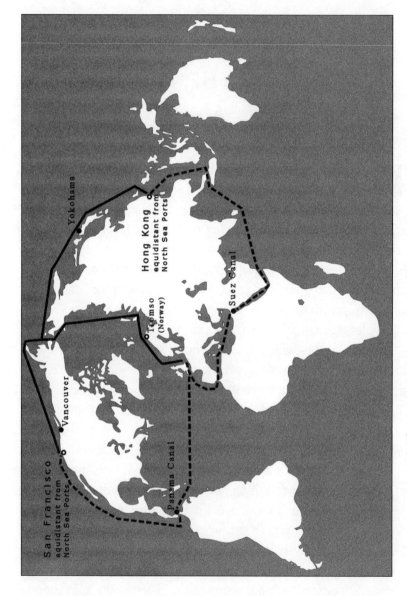

Map 10. The Arctic Option – potential distance savings by the arctic routes.

Having said all that, it is obvious enough that when starting a voyage from somewhere like Hammerfest in northern Norway, bound for Alaska or Kamchatka, the shortest route is across the Arctic Ocean. But what about a trip from, say, Rotterdam to the Japanese port of Yokohama?

At present, your ship will naturally take a southerly route – avoiding the ice. Yet Russians selling the prospects of the Northern Sea Route have for many years patiently explained that if you would only brave the cold, you could save almost 4,000 miles – that is more than a third of the 11,000-mile voyage by way of the Suez Canal. And there are still distance savings to be made through the Arctic even if your ship is eventually southbound beyond Yokohama to Shanghai. In fact, the point of equidistance as between the two routes is even further south, somewhere near Hong Kong.

A similar geography applies to the North American continent, though not quite so dramatically. Setting out from London for the Pacific West Coast, San Francisco is approximately equidistant by the two alternative routes – Panama Canal or North-West Passage. If your destination is anywhere north of that – Vancouver, for example – the arctic route wins. And as you move the departure point northwards into Scandinavia, the advantage obviously grows accordingly. An arctic voyage between the Norwegian port of Tromso and Vancouver would save more than 3,000 miles when compared with the Panama route.

A ship owner looking at these numbers will immediately start converting miles into hours at whatever speed his vessels can manage, and calculate from that the number of days saved on a given voyage. At 20 knots, for example, the Rotterdam-Yokohama reduction of 4,000 miles converts to 200 hours – more than eight days. A big vessel may burn as much as 200 tons of heavy fuel oil a day, so at several hundred US dollars a ton, there is a great deal of money to be saved even if the gain in time is not in itself critical. When oil prices shot up in 2008, at least one container ship operator ran its vessels several knots slower to conserve fuel, even though it meant inserting an extra ship to maintain a weekly service. Moreover an arctic routeing means there are no Suez Canal tolls (which for one of today's vast container ships would amount to several hundred thousand US dollars), and what is more, unless you count the mysterious hijacking of the Russian timber carrier *Arctic Sea* in August 2009, there are no pirates.

Piracy has never been completely eradicated, but since 2008, it has established itself in the Indian Ocean as a big business, a seemingly permanent factor in shipping calculations. As I write this, seven ships are reported anchored off the Somali coast, waiting for their owners to negotiate ransom demands running into many millions of dollars. Even a huge 300,000 ton tanker is evidently not immune from attack when deep-laden with crude oil, her decks almost awash. Insurance rates have gone up as a result, and some ships are being re-routed round the Cape of Good Hope, perhaps adding thousands of miles to their voyages. For the moment, the multinational naval task force assembled to combat the pirates appears almost useless, unable to intervene for fear of exacerbating the problem.

If the arctic routes had already been established, this situation would certainly have given them a boost. But since for most ship owners, the North-West and North-East passages are no more than a speculative prospect, bedevilled by unknown quantities, the relevant cost comparisons – in this case involving insurance rates as well as distances – are difficult if not impossible to make. Until the international marine insurance market has more experience of arctic conditions, voyages will be assessed more or less on a case-by-case basis. That is likely to be expensive, but just *how* expensive?

Changing course

The first thing any insurer will ask is whether the ship is ice-strengthened and otherwise appropriately equipped for arctic navigation – like the Russian and Finnish vessels specifically designed for those waters. A normal vessel may be able to cope in certain areas at certain times of the year, but some degree of hardening will usually be desirable, often essential. And that could add anything from 10 to 50 per cent to the cost of construction, something only justified if a ship is frequently deployed in the North. There is far more to an arctic crossing, in other words, than simply changing course.

While a northern route may well be substantially shorter, the real time saving could be reduced by the need to slow down or deviate to avoid heavy ice. And the mere fact that the duration of a voyage is uncertain will be a problem in trades where regular scheduling is important. Container ships shuttling back and forth between North-West Europe and the Far East – typically in recent years exchanging cheap manufactured goods for various forms of waste – prefer if possible to offer customers a reliable weekly service, each ship making a round trip of about eight weeks. Slotting in a few arctic crossings without disrupting that pattern would not be easy, quite apart from the fact that a direct northern routeing removes the possibility of intermediate calls in the Middle East, India or Indonesia.

The Russian authorities promoting the Northern Sea Route obviously understand all this; doing something about it is another matter. Their response is to keep plugging away at the theoretical advantages of the North-East Passage, promise to keep icebreaking pilotage tariffs as low and flexible as possible and do their best to get the funding needed to upgrade their marine infrastructure.

The latest simulation by the St. Petersburg research team compares the time taken by a 'traditional' container ship (i.e. a fairly small, ice-strengthened vessel carrying 2,500 standard containers) on a Suez voyage between Yokohama and Hamburg, with three different kinds of arctic transit – unescorted, escorted by the icebreaker *50 Let Pobedy* or accompanied by one of the new icebreakers soon to enter service – in various ice conditions.

Under 'light' ice conditions showing the effects of global warming, an escorted ship saves up to ten days or more during the late summer and autumn months by using the North-East Passage, and there are smaller savings even in winter. The unescorted ship's arctic transit takes substantially longer than a Suez voyage during the three late winter months but still improves on the southern route throughout summer and autumn. In other words, if the melt continues, strengthened ships will nearly always save some time by switching from Suez provided they hire an icebreaker, whereas operating independently, they will only gain for eight or nine months of the year.

In 'medium' ice conditions, the improved performance of the new icebreakers becomes significant and the unescorted ship is at a major disadvantage during the winter, only improving on Suez for six months of the year. But the escorted ships save some time for eight or nine months by going north-about, gaining about ten days on a voyage during the late summer.

The same analysis has been extended to estimate the cost of shipping a single container along the same Yokohama-Hamburg route. Under 'light' ice conditions, using the services of the *50 Let Pobedy*, the average cost is lower than for Suez during the arctic summer but higher during the winter, so that over a whole year there is not much in it. Independent operations produce a lower average cost on the northern route than through Suez, but without an icebreaker escort, a container ship cannot offer a year-round scheduled service.

One thing seems clear from this. There is a critical trade-off between the cost of icebreaking assistance and the more reliable transit times it provides. And either way, the level at which the fees are pitched, and the flexibility of the Russian tariff, will be crucial to the international success of the Northern Sea Route – just as the *INSROP* research suggested.

Reality check

In the summer of 2009, such simulations were finally put to a practical test as a pair of German freighters made what were claimed as the first 'truly commercial' east–west transits of the whole North-East Passage. The two vessels carried power plant equipment from South Korea to the Ob estuary, then sailed on to Arkhangelsk to load steel piping for West Africa. And from the Bremen-based Beluga Shipping Group's perspective, the whole exercise seems to have been thoroughly successful – and profitable.

Beluga is a relative newcomer to the shipping scene, with a rapidly expanding fleet of specialised heavy-lift cargo vessels equipped with their own powerful cranes, so as to be independent of normal

8. View from the bridge of *Beluga Fraternity*, navigating the North-East Passage, September 2009.

port facilities. And the ships chartered by General Electric for this job, *Beluga Fraternity* and *Beluga Foresight*, are both strengthened to work in ice. What is more, the German company was able to use two Russian captains – already on its payroll – to make the pioneering voyage.

Their cargo – 44 large chunks of power plant, each weighing 200–300 tons, destined for a power station in the Siberian city of Surgut, in western Siberia – was loaded at Ulsan, South Korea, towards the end of July. After compulsory Russian inward clearance at Vladivostok, they headed through the Bering Strait and across the Northern Sea Route, using the Vilkitsky Strait – often an icy chokepoint – to enter the Kara Sea and the Ob river estuary. At Novy Port, hundreds of miles inland, and hampered by bad weather, they used their own gear to transfer the heavy cargo into barges. From there they would have headed for Murmansk, western gateway of the North-East Passage, for outward clearance by the Russian authorities, but at the last moment Beluga was able to arrange a second cargo for each ship – 6,000 tons of steel piping – from Arkhangelsk to Onne, Nigeria. By September 18, they were both on their way south.

Across the central section of the arctic route, the German freighters were escorted by nuclear-powered Russian icebreakers – one each, the *50 Let Pobedy* and the *Rossia*. The big black-hulled icebreakers shepherded them through the middle of the Novosibirskiye islands, using the Sannikov Strait, across the Laptev Sea and through the Vilkitsky Strait into the Kara Sea.

Hidden beauty

After all the anticipation, not much ice was actually encountered. The worst of it, as expected, was in the Vilkitsky Strait, where at one point ice half covered the sea, but nothing that worried the Russian skippers. Captain Aleksander Antonov, aboard the *Beluga Fraternity*, seemed more concerned about the frequent fog, obscuring the beauty of the small icebergs they were passing.

His company said afterwards that its ice-hardened ships could have managed perfectly well without icebreaker assistance, but the Russian authorities made it a condition of their granting permission for the voyages. In any case, Beluga was happy to accept this extra security because the customer wanted it – and picked up the icebreakers' bill. The Russian authorities' decision on whether an icebreaker is required depends on the ship's ice classification, what route it takes and in what season. In the Vilkitsky Strait – defined as 'internal waters', and blocked by heavy ice for much of the year – icebreaking support is compulsory.

Although ice was not a problem on this particular trip, Captain Antonov said he nevertheless found navigating the unfamiliar route 'a challenge'. He had to deal with the various port and coastguard authorities – for whom being able to communicate in Russian was obviously most helpful. And as he pointed out, on another occasion they might encounter more ice.

For his colleague Captain Valeriy Durov, aboard *Beluga Foresight*, the most difficult part of the trip was unloading the heavy cargo at Novy Port offshore, rather than alongside a quay – 'because discharging heavy lifts offshore is always challenging, and when we had bad weather we had to stop, and wait two more days to complete the operation'. But then that was the sort of operation for which these specialised *Beluga* ships were designed.

During the crossing, both captains used radio interviews to extol the benefits of arctic routeing while showing an appropriate concern for the environment. Antonov credited global warming with opening up the passage, enabling them to save time and fuel, but added tactfully that: 'Personally, I think we have to be careful with our planet and the environment in general, not only with regard to the North-East Passage and the melting ice.'

In Bremen, his enthusiastic boss, Beluga Shipping's chief executive Niels Stolberg, predicted that more voyages would follow this first effort. Those companies with suitable ships would in future make quite regular use of the new seaway, at least during the summer. But he acknowledged the wider dangers of global warming, expressing the hope that the Arctic Ocean's summer melt would continue to be followed each year by a winter freeze. Meanwhile shorter voyages meant burning less fuel, which in turn meant less harmful emissions.

From a commercial point of view, the voyage costings indicated by the Bremen firm look encouraging. Each ship apparently saved about ten days by taking the Northern Sea Route instead of the long haul south-about through the Suez Canal, thereby saving 200 tons of fuel (20 tons a day) at $500 a ton – a total saving on fuel of $100,000. Other operating costs amounted to about $20,000 a day, producing a notional ten-day saving of $200,000. Each ship therefore saved a total of $300,000 by taking the arctic short cut.

This is not of course a complete financial calculation. Most importantly, it does not include the mandatory icebreaker fees paid in this case by the customer. But Beluga's remarkable pioneering effort nevertheless illustrates exactly the sort of foreign shipping a successfully reformed Northern Sea Route authority could expect to attract in the coming years, given a bit of help from global warming – an even better prospect if the voyage could be done independently.

MELTDOWN

There is every sign that help from the climate will be forthcoming – good news for the maritime community but not a prospect welcomed by scientists now anxiously monitoring the Arctic. They fear that as the icy cauldron of the Arctic Ocean melts, the changes that occur – both above and below the ocean surface – may well have a profound effect on the rest of the world's climate. And for many of us, the effects could be far from benign. The open, ice-free water that sailors celebrate is the last thing climatologists want to see. It is the harbinger of potentially disastrous global warming.

The Arctic is an extreme environment which amplifies climatic change. The central effect, amid a sea of complexities scientists do not pretend fully to understand, is that temperatures there are rising almost twice as fast as elsewhere on the planet. The Arctic Climate Impact Assessment, a four-year study involving hundreds of experts which reported in 2004, projected a 4–7 degrees C rise by the end of the century.

That may not sound much, given that we experience far bigger seasonal variations every year, but as an average change, it is enough to produce dramatic effects. Sea ice will increasingly melt, endangering the ringed seals that depend on it and the polar bears that depend on hunting the seals. Glaciers will retreat, exacerbating a rise in sea levels that threatens low-lying areas from Louisiana to the Maldives. The arctic forests will creep northwards – a

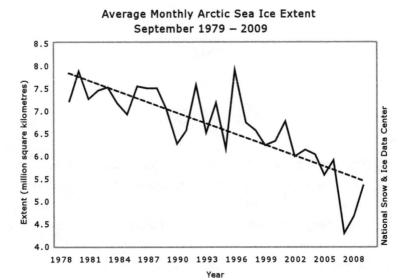

Average Monthly Arctic Sea Ice Extent
September 1979 – 2009

beneficial change in that trees absorb carbon dioxide, but pushing back the tundra on which indigenous reindeer herders depend. The tundra's underlying permafrost – permanently frozen soil – will thaw, threatening arctic roads, pipelines and buildings with collapse and releasing methane far more harmful than carbon dioxide as a 'greenhouse' gas. Some fisheries, such as cod and herring, are expected to benefit, but here as elsewhere the forecast changes are complex and therefore uncertain.

The most visible of these effects is the vanishing ice which prompted this book. Since 1979, when space satellites made accurate records possible, the sea ice surrounding the Pole has gradually shrunk. Each year in March it reaches a maximum 'extent' (a technically defined term that includes the ice-strewn fringes) and a minimum, after a summer time lag, in September. Over the past 30 years, although there have been a number of exceptional seasons, with much more or less than the average cover, the minimum extent has decreased by about 30 per cent.

In September 2002, the ice shrank to a record minimum. Another record low occurred in 2005, followed by a slight recovery the following September. Then in 2007, the downward trend suddenly accelerated. That autumn the ice covered only just over 4 million square kilometres – more than a 40 per cent reduction from the 1978–2000 average summer minimum extent of about 7 million square kilometres. More than a million square kilometres, an area about five times the size of the United Kingdom, had been lost in just two years – or from the other perspective, gained as clear ice-free water.

Ice floes still covered the North Pole, but on the eastern side, the ice edge was at the highest latitude ever recorded – 85.5 degrees N. If Nansen had waited long enough, he could have sailed to within about 300 miles of the Pole before starting his frozen drift. Thicker

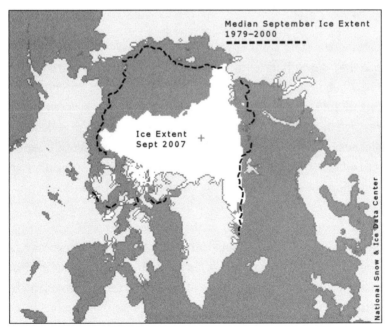

Map 11. 2007 – A Record Year for the Summer Ice Melt.

multiyear ice remained piled up along the Canadian and Greenland shores, but large stretches of the Siberian coast were ice-free and more remarkably, the main deep-water North-West Passage – the Parry Channel – was almost totally clear.

Satellite images of the Arctic Ocean in mid-September that year, 2007, showed the residual pack ice in a roughly triangular shape. One corner reached into the Beaufort Sea, though well clear of the mainland; another trailed down the eastern side of Greenland; and a third clung to the islands of Severnaya Zemlya, at the northernmost point of the Siberian coast. On either side, the Kara Sea (its 'ice cellar' reputation notwithstanding) and the Laptev Sea were open. In maritime terms, this was a startlingly good year for the North-West Passage but not so good for the North-East Passage because of the blockage at the Vilkitsky Strait.

Alarm call

It was in that late summer of 2007 that the world woke up to the fact that something extraordinary was happening in the Arctic. It had been happening for many years before that of course, but stark pictures of a shrinking ice cap, taken from space, conveyed the message in a way that graphs of rising temperatures or reduced ice thickness did not. Journalists hurried to question scientific experts as to what it might mean. Scientists responded with some alarming predictions of an ice-free ocean. The polar bear became an endangered species.

Accordingly, the progress of the subsequent, 2008, season's ice melt was followed with intense interest. Larger areas of exposed sea were able to absorb the sunlight's warmth, attacking the ice from below, while higher than average atmospheric temperatures melted it from above. In the event, the ice cover shrank not quite to the previous year's spectacular minimum, but nonetheless well below the long-term trend, which continued downward.

It was on August 27, 2008, that Mark Serreze, director of the National Snow and Ice Data Center in Colorado, pointed out that the North-East and North-West passages were *both* open for the first time in modern recorded history (i.e. since the 1950s). No one knows when it last happened, but it was probably a long time ago. It has even been suggested that this was the first such coincidence since the last warm interglacial period more than 100,000 years ago.

The other important feature of the 2008 arctic summer is that although the extent of the ice cover was much the same as in the previous year, there was a lot more thin first-year ice and correspondingly less of the difficult multiyear ice – collected mainly along the Canadian and Greenland shores. The total volume remaining to keep the ocean cold was probably smaller even than in 2007, just spread more thinly.

In 2009, the September minimum ice extent was the third lowest on record, showing another modest recovery from the exceptional situation in 2007, but not enough seriously to disrupt the downward trend in coverage and thickness. The NSIDC calculated that summer arctic sea ice coverage, as measured each September, was by now declining at a rate of 11 per cent a decade. The centre's director Mark Serreze commented that it was nice to see a small recovery, but there was no reason to suppose we were heading back to conditions seen in the 1970s – 'We still expect to see ice-free summers sometime in the next few decades.'

During 2009, the pack ice retreated in different areas because of changing wind patterns. As in the previous year, both arctic coastal passages were briefly open to shipping. But whereas in 2007 the main Parry Channel through the Canadian archipelago had been free of ice – the channel that defeated Queen Victoria's navy and gave the icebreaking tanker *Manhattan* such a hard time – in the two following seasons only the shallow southern channel opened up – the route which trapped Franklin and was later painfully negotiated by Amundsen.

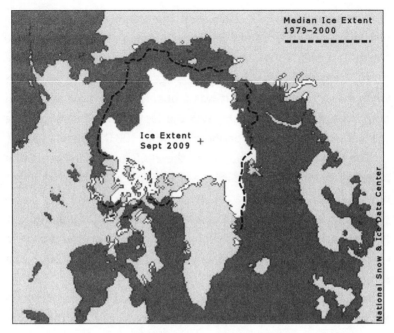

Map 12. Summer Ice Extent – 2009.

A record number of vessels used the North-West Passage during the summer of 2008, perhaps prompted by the previous season's exceptionally clear water. But there are compelling reasons to think that even if the ice continues to melt as predicted, the Canadian route will not be *reliably* open – which is what matters to shippers planning cargo movements – for some time after its Russian counterpart, the Northern Sea Route.

It is already clear that what remains of the thicker, multiyear ice is concentrated on the Canadian side of the ocean. While it lasts, there is always the danger of its flowing southwards, driven by the prevailing winds to block the archipelago's channels – indeed increasingly so as the pack ice breaks up. It was this pressure which bedevilled the efforts of the nineteenth-century British explorers and finally destroyed Franklin's expedition. There was a let-up in 2007, and to dramatic effect, but by 2009 the

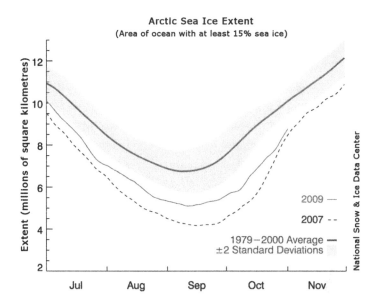

Canadian Ice Service was reporting abnormal concentrations of thicker, multiyear ice in the central section of the Parry Channel, complicating its navigation, because of an earlier surge of older ice from the Beaufort Sea.

If this pattern continues – that is, slow shrinkage of the summer ice cap from the Siberian side, leaving awkward drifting chunks to block the Canadian archipelago – it may well be that not only the North-East Passage but also a central Trans-Polar Route will be open to regular shipping before the North-West Passage. That is indeed what the Canadian Ice Service and others expect to be the order of events.

Not that anything about the arctic climate is easily predictable – as the winter of 2009–10 reminded us. Notwithstanding the records collected by pioneers like Nansen, by the Soviet drifting ice stations or US Navy submarines, the necessary data are incomplete – inevitably so, given the hostile environment. Until aircraft were available – initially just in the clumsy form of airships – nobody could be absolutely sure what was there. As we have seen, a complete group of islands off the Siberian coast – Severnaya Zemlya – went unnoticed until 1913!

The advent of surveillance satellites in 1979 transformed the situation, at least as far as the icy surface is concerned. Cores extracted from deep in the Greenland ice cap have also revealed a hidden climatic history reaching back thousands of years. Submarine navigators and oil company geologists have helped fill in some of the gaps while pursuing their own agendas, and nowadays the prospect of radical climate change is of course prompting direct research on a large scale. Yet even where substantial information is available, and even with the help of powerful modern computers, prediction is an horrendously difficult exercise.

Abruptness

In a mathematical sense, the process of maritime climate change tends towards the chaotic. Different temperatures, densities, winds and currents constantly mingle to produce the weather which defines a particular climate. In Britain, for example, we live roughly on the 'polar front' that divides the cold air masses of the northern regions from the warm air of the tropics, and the convective pressure waves that swirl along it give us our characteristic weather 'depressions' – warm rain followed by cold blustery showers and bright periods. Like the rest of North-West Europe, we also benefit from the warming effect of oceanic convection in the form of the Gulf Stream. The Arctic Ocean has its own deep currents, and above the ice, weather systems which can force cold air outwards, or suck warm air in, block the sunlight with cloud or clear the sky to let it through.

These elements constantly interact. One thing leads to another, but not just directly. There are often feedback mechanisms, like the one described earlier that seems to drive the sea ice melt. And some of these may have the potential to accelerate to one of those dangerous tipping points, producing what scientists refer to as 'abrupt climate change' – an academic's understated description of a potentially devastating event.

Take the latent problem of methane, 20 times more effective as a 'greenhouse' gas than the carbon dioxide whose rapid build-up in the atmosphere is arousing so much concern. Huge quantities of this unpleasant gas are trapped not far beneath the Arctic, in icy sea bed structures or buried in the frozen subsoil – the permafrost – of the surrounding tundra.

Demonstrating the presence of methane is surprisingly easy if you know where to look. The American researcher Dr Katey Walter, from the University of Alaska Fairbanks, performs a spectacular scientific party trick by drilling a hole in a frozen lake, then igniting the methane jet it releases. At sea, numerous plumes of methane have recently been found along the Siberian coast and off the west coast of Spitsbergen, apparently bubbling up from perforations in the frozen sea bed.

Work is under way to determine how serious a phenomenon this may be, and on what timescale. There is no doubt that the permafrost is beginning to melt, posing a serious problem for arctic operations by threatening buildings, roads and pipelines constructed on what was previously solid frozen ground. The deeper fear is that as the warmth spreads, a massive methane release could trigger runaway change, stoking up the greenhouse effect to provoke further releases in a vicious cycle. There is evidence that this may have happened in the past, in one instance perhaps being implicated in a mass extinction of living creatures.

From meltdown to shutdown?

Another alarming scenario associated with melting arctic ice is possible disruption of the ocean current warming North-West Europe – the Gulf Stream that prevents Britain sharing Labrador's winter cold and keeps the port of Murmansk ice-free. And again, this is something that has apparently happened before, albeit in different conditions, at the end of the last ice age. About 13,000 years ago, a vast lake of meltwater that had accumulated near the modern Hudson Bay burst out into the Atlantic, halting, or at any rate helping to disrupt, the warm current. According to Bill Patterson of Saskatchewan University, who found a record of the event in Irish lake sediments, it took only a few months. As he put it, it was suddenly as if 'Ireland moved to Svalbard'.

The Gulf Stream is just one branch of a system of oceanic currents that snake round the world and back again, driven by winds, water temperature and density. (Its internal mechanism is often referred to as the 'thermohaline circulation', i.e. dependent on heat and salinity.) The section that concerns us here is the warm, surface current running north-east from the Caribbean, past the British Isles and round the North Cape of Norway into the Arctic Ocean.

As the stream moves north, it sheds its heat, becoming saltier and denser and therefore eventually sinks to the ocean bottom where it drifts southwards to balance the north-going current on the surface. This benign circulation is now threatened in several ways – by reduced ice formation (because when the sea freezes, it sheds its salt into the surrounding water), by fresh water pouring off the Greenland ice cap as it thaws and by the probability of increased rainfall. And there are signs that these forces are taking effect.

The sea in this corner of the North Atlantic has become less salty. The cold water flowing through the Greenland–UK gap has slowed. And most dramatically, on one of his recent submarine trips, Professor Wadhams found that the vast underwater 'chimneys' through which water previously sank to the sea bed beneath the Greenland ice shelf had almost disappeared. There does seem to be a serious possibility that the Gulf Stream could weaken, or even shut down altogether.

If that happened, the climate of Britain and Norway in particular would be transformed – becoming significantly cooler – and the indirect effects would spread right round the globe, some of them decidedly unpleasant. As already mentioned, the climatological record exposed in Greenland ice cores and ocean sediments indicates that the Gulf Stream has slowed and stopped altogether in the past – not at the cataclysmic speed suggested in the film *The Day after Tomorrow*, but fast enough to disrupt our present ways of living, particularly agriculture, right along the coasts of North-West Europe – and incidentally deprive Murmansk of its historic ice-free status. In that case some of us might welcome a little global warming to offset the arctic cold.

POSSIBLE OUTCOMES

Before entering the maze of possible outcomes from the arctic melt, there are a few that can usefully be excluded. For a start, the ice covering the Arctic Ocean is not going to disappear overnight – from which it follows that any resulting changes in maritime trading patterns will be gradual, not sudden and dramatic.

Although the 2007 retreat did prompt some startlingly early predictions of an ice-free summer – one suggested it might happen by 2013 – scientists now seem more inclined to hedge their bets. The downward trend in ice cover and thickness is clear, and it might well accelerate, but everyone accepts that individual years and individual areas – like the North-West Passage – may buck the trend. Russian prognoses in particular often include a warning from 'some experts' that the warming process is cyclical because – if for no other reason – that is their experience over long years of struggling with the North-East Passage. And the unusual weather patterns of the 2009–10 winter seemed to bear that out.

Even assuming the ice *does* clear quickly, perhaps quite suddenly, it will only happen during the short arctic summer. That will certainly make things easier for the sailors and oilmen while it lasts – perhaps eventually months rather than weeks – but apart from occasional visitors, it will not enable them to dispense with their expensive ice-strengthened ships and low-temperature equipment. Some expensive icebreaker assistance will still be needed. Difficult ice will persist in some areas, particularly along Canadian shores – and where there is less ice, it may be dangerously on the move. Dark, cold winter conditions will return each year.

This obvious fact is especially pertinent where international traffic between the Atlantic and the Pacific is concerned. It is one thing to send existing ships like the two Beluga vessels on a one-off voyage through the northern route, quite another to invest in new ships whose expensive modifications can only pay their way for a few months of the year. And that is before taking account of the numerous other factors that come into play – draft limitations, inadequate charts, insurance costs, icebreaker support fees, political uncertainties and so on.

For deep-sea sailors in general, there is a broad comfort zone between latitudes 60 degrees north and south, within which communications are reliable and icebergs are unlikely to spring a nasty surprise. Persuading them to leave it will not be a simple matter. In short, the disappearance of summer ice will not prompt a maritime revolution, instantly transcending the geographical advantage of the Suez Canal, so much as a process of piecemeal evolutionary change.

Double negative

Nor is the North Pole going to be the setting for a new kind of Cold War – much as it might make for an easy headline. While it is true that the politics of warfare are often wildly unpredictable (Until Bush and Blair got together in the aftermath of 9/11, for example, who could have predicted that Britain would rapidly be drawn into two long, brutal and debilitating wars?) there are solid reasons why the main players in the arctic arena should avoid confrontation.

There are admittedly numerous conflicts of interest, not excluding close allies like Canada and the USA, who have radically different views on the legal status of the North-West Passage and are in dispute over their shared continental shelf. There has even been some gentle sabre rattling as Canada asserts its sovereignty and Russia rebuilds its naval strength. But this does not mean they are looking for trouble. Moscow's immediate overriding priority must surely be to strengthen its economy, whether by reforming its 'primitive' centralised structure, as President Medvedev appears to advocate, or by simply exploiting its abundant natural resources.

Either way, the most obvious new source of economic prosperity is in the gas, oil and rare minerals hidden beneath the Arctic. Finding, extracting and selling them at a profit is a lengthy process, measured not just in years but in decades. It can only be achieved from a reasonably stable platform of international technical, legal and financial co-operation – international oil companies want to know where they are going to pay their taxes. The Kremlin surely knows this, even if Chilingarov does not. And that suggests that expensively suited lawyers, rather than men in uniform, will be in the front line of any arctic conflict.

In 2009, the Arctic Council's *Arctic Marine Shipping Assessment (AMSA)* tried to make sense of the future by charting a quadrant of possible scenarios, with one axis measuring the extent of trade in the region's resources and the other the extent to which it will be governed by stable rules of conduct. If economic demand were high and governance merely responded ad hoc to events, the stage would be set for a 'race' to grab a share of the spoils, whereas stable rules-based governance would allow healthy development while showing concern for the Arctic's indigenous cultures and vulnerable ecosystems. Low demand could lead to a murky, unstable and underdeveloped future or, given the right controls, to gradual development with extensive room for environmental protection, including possible 'no-shipping zones'.

A simpler approach is to consider what events might inhibit or foster progress towards what I take to be a widely agreed objective – orderly exploitation of the region's economic resources, involving heavy reliance on marine transport, without doing further damage to its indigenous communities or soiling its pristine physical environment.

Dangerous waters

There are two accidents waiting to happen – a major offshore oil spill and a serious emergency involving a passenger ship – which, while not neutralising the powerful forces propelling the Arctic's development, might nevertheless shape its course.

A mass of black oil staining the ice – catching more public attention than the 2006 Alaskan pipeline leak – would challenge industrial claims to have learned the lessons of past experience. Response mechanisms developed in anticipation of fresh offshore operations would be put to the test. If the spill involved a ship, as opposed to a pipeline or a drilling rig, it might also prod the maritime authorities into filling gaps in the regulatory barriers painstakingly erected to prevent some ill-equipped 'rust bucket' entering the vulnerable arctic zone – uniform standards of construction, inspection and training, with mandatory ship reporting along Canadian and Russian lines and so on.

Television film of oiled seabirds and blackened rocks have made us all too familiar with the devastation a ruptured *Torrey Canyon* or *Amoco Cadiz* can cause. But those casualties did at least occur on coasts exposed to pounding, cleansing Atlantic breakers. A combination of oil and ice presents even greater problems. The oil will lie on top – virtually impossible to collect and difficult to burn – or be trapped beneath, where its presence, perhaps leaking from a damaged pipeline, might not even be detected until the ice thaws and disperses.

The maritime community is of course well aware of the threat. Emergency training exercises have been organised – for example, at the new Varandey offshore terminal in the Barents Sea, where tankers now load surrounded by ice. A multinational exercise involving a simulated collision between a shuttle tanker and a vast storage vessel – one of an annual series jointly staged by Finland, Norway, Russia and Sweden – was held at Murmansk in 2009.

However, there is an old military adage that no plan survives contact with reality. And even if in this case practice makes perfect, as one obviously hopes, the real event would come as a shock, with widespread political fallout. PR departments would be working overtime. This time it could be pictures of an oiled polar bear on our television screens, ramming home the message that oil and ice do not mix.

A serious emergency involving a big cruise liner carrying thousands of mainly elderly passengers (as opposed to the small vessel which sank in the Antarctic) would also give the system a salutary jolt. Dozens of passenger vessels nowadays head north each summer. Some of them are icebreakers, with a few adventurous customers on board. Others are large conventional cruise ships diverted from their usual Caribbean haunts to take a look at polar bears and calving icebergs. No study of the Arctic's maritime future ends without a warning that search and rescue facilities are few and far between (because until now there has been no shipping traffic to justify the cost of dedicated aircraft, tugs and so on) and that these waters hold special dangers, however much care is taken.

Setting such anxieties aside, one can list perhaps ten factors that will determine how far and how fast the arctic maritime scene is going to change:

The price of oil and gas, notably volatile of late, is an important determinant because the Arctic's reserves will always be an expensive option. For large parts of the year, this will remain a dreadful place to work – bitterly cold, dark and stormy. That inevitably impacts on the costs of exploration, extraction and transport when compared with the Middle East, America or wherever else. If it can be avoided, it will be.

9. Will this container ship leaving Felixstowe one day head North through the Arctic, not South through Suez?

While physical conditions are prohibitive right across the Arctic, activity in the Canadian and Alaskan sectors is particularly subject to **political argument** over the rights of indigenous communities and the need to protect their vulnerable environment from the demands of our industrialised appetites. Here is a clash between Sarah Palin's 'Drill, baby, drill' mentality and those who believe a wilderness like the US Arctic National Wildlife Refuge has its own value, and should remain sacrosanct. At what point – if at all – should local people be expected to stop hunting whales or caribou, and consign themselves to a company job? And if the hunting stops, do the whales then lose the limited protection they currently enjoy? Do they too have rights?

Further exploration in this region, especially if it takes place offshore, raises another question directly impacting on maritime activity. Will retreating ice prompt **a reversal of the strategic decision** represented by the costly construction of the Trans-Alaska Pipeline in the 1970s, so as to allow the use of oil and gas tankers?

The outcome will depend on production volumes and geography as well as the possibility of a major oil spill, but this issue will surely be revisited by the accountants.

The **legal title** to all that potential wealth buried beneath the arctic basin has yet to be finally established. As explained earlier, there is a cat's cradle of overlapping claims under the *UN Convention on the Law of the Sea (UNCLOS)*. When the 2014 deadline for the last of the currently scheduled bids arrives, the legal fur will no doubt fly. But since, as of now, no US administration has yet managed to persuade Congress to ratify the UN treaty (even though it would allow Washington to lay claim to 45,000 square miles of continental shelf), it could be many years beyond that before the final boundaries are drawn.

While demand for energy drives shipping within the arctic zone – the Shtokman gas field alone is expected to keep 30 ships in business – transit traffic, such as it is, will be determined by how **world trading patterns** settle down after the current recession. More imponderables, in other words, though it will certainly take a lot to persuade those big container ships to change their routine.

The future of **the Russian economy** is particularly obscure, because it depends so much on political rivalries yet to be resolved – and barely understood by outsiders. And the **Middle East**, the predominant source of our present oil supplies, is of course horrendously unstable. In the past, this led to lengthy closure of the Suez Canal as British, Egyptian and Israeli troops fought over it, though the chances of its doing the arctic routes that kind of favour in the foreseeable future are remote. For the moment the desert waterway remains, along with Panama, merely the yardstick against which the theoretical savings of an arctic transit are measured.

Beluga's pioneering voyages suggest it may soon be more than a theoretical exercise. Ship owners may, at least occasionally, compare the level of **canal tolls with** the fees charged for hiring an icebreaker before deciding whether to use the Northern Sea Route. However, many other things must first be resolved.

Most fundamental, of course, is how soon the Arctic Ocean's ice-free summer actually arrives. You can still take your pick from forecasts ranging across much of the coming century, but I would not be writing this speculative book if I did not myself believe the trend is clear, and likely to be permanent.

Almost equally important is the way the ice disperses, and the weather that accompanies its break-up. Assuming Canadians are right that extremely variable ice conditions in the North-West Passage mean it will be the last route to open as a reliable international waterway, transit shipping will for years be at the mercy of Russian bureaucracy and Russian charts. Unless, that is, the central ice pack rapidly retreats to reveal a deep-water Trans-Polar Route straight across the central ocean from the Greenland Sea to the Bering Strait, from the Atlantic to the Pacific. In which case, perhaps navigators can just set the 'satnav' and go.

Yet this still begs the question of what an 'open' sea lane means in these extreme latitudes. How many dangerous chunks of ice will still be drifting around? How much strengthening will your ship require? At what point will it be worth building a specialised, ice-hardened arctic cruise ship? What speeds can safely be maintained and schedules guaranteed? The skippers of refrigerated gas tankers, for example, like to crack on at high speed to reduce the amount of cargo boiled off in transit – and it seems there will soon be a lot of these ships loading in the Arctic.

Shipping forecast

And what will the weather be like? When even the familiar patterns that dictate Britain's weather occasionally defeat the Met Office, simulating what may happen in a radically new arctic scenario is not at all easy. From a sailor's point of view, the outlook is discouraging.

One recent study reported from Norway predicted 'large increases in the potential for extreme weather events' right around the rim of the arctic basin as the ice cover shrinks. While calm conditions often prevail above the ice cap itself, the seas around it are prone to phenomena known as 'polar lows' and 'arctic fronts'. Cold air leaving the ice can be lifted into an explosive storm.

Another warning came from Alexander Frolov of the Russian state weather forecasting organisation *Rosgidromet*. His concern was that drifting icebergs and more frequent ice storms could threaten development of the Shtokman offshore gas field in the Barents Sea, which unlike Norway's amazingly automated Snohvit field is expected to use floating platforms.

Then there is fog, already a persistent problem for arctic navigators. Nothing scientific here – just my own hunch based on the basic fact that fog forms when air cools below its 'dew point' and sheds its moisture (which is why we have to demist a cold windscreen when starting the car on a winter's morning). There is plenty of scope for this phenomenon as the boundary between cold ice and warm sea shifts. So perhaps the polar region's typically poor visibility will get even worse as the arctic routes make their seasonal transition.

In any case, satellite navigation using the global positioning system is going to be at a premium. GPS is based on a network of US military satellites (rival Russian and European networks are under development) which can be constantly interrogated to give a position on the globe accurate to within a few metres. It was in use at sea before it appeared in cars as a route finder, and is now integrated with electronic charts, and used to send out automated distress calls. The technology is considered so reliable you could legally take a ship to sea without any paper charts on board – a fact that still shocks me. Once full, accurate coverage is available, GPS will be absolutely invaluable in a region where traditional navigational marks are scarce and radar is sometimes suspect.

Underlying all this are the desperate uncertainties of **global warming**, where public debate takes in everything from crude conspiracy theory to serious scientific explanations which nevertheless assign quite different values to factors such as human-made carbon dioxide or cyclical solar activity. Meanwhile the climate-change sceptics are still in full cry. So in a discussion like this, one needs to put one's cards on the table.

From my layman's perspective, there is thoroughly convincing evidence that we are heading for a critically dangerous temperature rise of several degrees centigrade by the end of the century unless governments take drastic action. I am told there is a mathematical concept known as 'catastrophe theory' describing the way normal, apparently stable processes suddenly run out of control. It seems to me our climate is deeply prone to that. We could indeed be on the brink of a disastrous tipping point, and I fear our essentially short-term systems of government are ill-equipped to avert it.

Since none of this is certain, discussion of what it means for the Arctic can be no more than a plausible hypothesis; the only certainty is uncertainty. Yet the arctic basin is still the one to watch – the amplifier of climate change and the source of so much hidden wealth, disputed legally even if it never results in armed conflict. Russia, if she plays her cards right, will be among the main beneficiaries; arctic wildlife, and the indigenous communities still dependent on it, will be among the losers.

There will certainly be a big surge in maritime activity – it has already begun – mainly associated with the exploitation of oil, gas and minerals within the region, plus polar tourism. Only gradually, as 'ice blink' is replaced by a 'water sky', will ships in transit from the Atlantic to the Pacific begin to head north. The maritime tipping point may well arrive when it becomes profitable to operate an arctic container shuttle between regional hubs in, say, Iceland and Japan, using ice-strengthened, purpose-built ships capable of independent operation.

If the ghosts of Barents, Franklin, Nansen and the rest are watching, they must by now be wondering whether all that hardship and sacrifice was worthwhile. Mary Shelley, with her vested interest in a beautiful frozen wilderness, will be seriously put out.

Northern Poll

Having personally hesitated to make hard predictions, I was nevertheless keen to hear from some of the experts I consulted in preparing this arctic essay. I asked them for a brief answer to a deliberately simple, all-purpose question: **'In what important way will the Arctic change by 2040?'** Their responses, received between October and December 2009, and recorded here in alphabetical order, speak for themselves:

Professor David Barber, Canada Research Chair in Arctic System Science, University of Manitoba

By 2040, we will no longer have multiyear sea ice in the northern hemisphere. The arctic sea ice cover is changing rapidly. Recent research into multiyear ice (ice that survives the summer and grows the next year) shows that less than 20 per cent of the total arctic basin is now covered with this old type of sea ice; yet as recently as 2007, the basin was about 70 per cent covered by multiyear ice.

On a recent field expedition to the Arctic (in September 2009), my team and I discovered that what satellites thought was a recovery in the extent of thick multiyear sea ice at the end of the summer melt period was in fact a misclassification by the satellite sensors. What we found instead was very heavily decayed sea ice left over after multiyear ice had disintegrated. This was covered by a very thin (less than 5 cm) cover of new ice. The dramatic reduction in multiyear sea ice continues at a very fast rate in the high Arctic.

By 2040, we will have no ice cover at all in the summer months, and only thin and highly mobile first-year ice in the winter. The arctic basin will have many drilling platforms that support hydrocarbon development since by then we shall have exhausted many of the more southern reservoirs.

We shall also see a lot of trans-polar shipping traffic, most of it going straight across the Pole, serving as a short cut between Europe, North America and Asia.

The marine ecosystem will be more productive over the arctic continental shelves (due to increased light and nutrient upwelling onto the shelf) and the ecosystem will support animal species which in our time reside in the North Atlantic and the North Pacific. The charismatic megafauna (polar bears, seals and whales) currently found in our Arctic will be restricted to the most northerly latitudes where the open water season is the shortest and the sea ice and snow cover the thickest.

By 2040, the serious issue of rising sea level, caused by the melting of glaciers in the Arctic and temperate parts of the planet, will begin to be felt, and people will have to grapple seriously with how to adapt to the reality of global climate change.

Simon Bennett, Secretary, International Chamber of Shipping, London

The role of soothsayer, particularly with regard to an industry as volatile and unpredictable as international shipping, is always fraught with danger. However, if the scientists are correct and the Arctic is indeed ice-free throughout the summer months by 2040, this will obviously have an impact on the flow of maritime trade.

Clearly there are likely to be new opportunities for ports to open up in Canada and Russia, with new sources of crude oil and gas, as well as timber and manufactured goods, having a more direct route to their markets using ships as opposed to land-based transport travelling for thousands of miles through remote continental interiors. There could also be a significant expansion of the use of specialist offshore support vessels that might service an expanding oil and gas exploration industry facilitated by the retreat of the ice.

However, much will depend on the speed with which new shipping routes are able to be subjected to the necessary hydrographic surveys in order to make them safely navigable, and further technical improvements will need to be introduced with regard to the development of large 'ice class' ships – the hazards presented by ice are unlikely to be removed completely, even if the predictions of the scientists are correct. Given the strategic sensibilities that may prevail in the areas concerned, it is also quite possible that Arctic navigation may not be opened up fully for political and defence reasons.

Moreover, the current economic factors that determine world trade routes may possibly remain fundamentally unchanged. Oil production, notwithstanding any increased use of low-carbon alternatives, is likely to continue for decades and remain centred in the Middle East and Central Asia, where most remaining oil reserves currently lie. The manufacture of consumer goods is likely to continue to be concentrated in South East Asia; and the location of major population centres is unlikely to change dramatically. It is therefore probably unlikely that there will be the wholesale shift away from current maritime trade routes – by large container ships, tankers and bulk carriers using the Pacific, Indian and Atlantic oceans – that some commentators might anticipate.

However, the future is uncertain, and we must always be ready to be surprised.

Dr. Lawson W. Brigham, Professor of Geography and Arctic Policy, University of Alaska Fairbanks and Chair, Arctic Marine Shipping Assessment of the Arctic Council, 2005–9

By 2040, the Arctic will be more integrated with the global economy. This economic linkage will be driven by natural resource development (for example, oil and gas, nickel, high grade ore, zinc, coal, timber and more) in Canada, Greenland, Norway, Russia and the USA (Alaska). Marine transport systems will support development and carriage of these resources out of the Arctic to world markets. While the Arctic will be a warmer place by 2040, the sea ice cover will remain for 9-10 months annually, requiring the use of polar class ships. The sea ice will be thinner, there will be less extent, and there will be minimal multiyear ice remaining. Increasing marine access and potentially longer seasons of navigation will be possible.

The Arctic in 2040 will be a peaceful region, where conflict has been avoided and the arctic states are engaged in a broad range of co-operation. Co-operative law enforcement operations have taken place during the past three decades, but there has been no direct conflict between the arctic states.

Kimmo Juurmaa, Director, Concept Development and RDS, Deltamarin Contracting, Finland

The climate change will proceed, but the changes in the Arctic are not as dramatic as proposed. The ice conditions will have huge yearly variation, and they will be unpredictable. Technology in ice monitoring will develop to the level where ice information is part of the regular weather information. This will help marine operations in the Arctic.

Operations will increase, but they mainly relate to the development of natural resources. Shipping will be the mode of transporting oil, gas and minerals. Trans-arctic shipping will remain on an occasional basis. Safety regulations will make arctic shipping expensive. Tourism will increase, and special passenger vessels for arctic cruises will be built. The coastal states will base their rules and regulations on national, area-specific requirements. The military presence will increase.

Technology in icebreaking will not make any giant leap, but there will be more independently operating vessels as the multiyear ice diminishes.

Dr. V. I. Peresypkin, Director, Central Marine Research and Design Institute, St. Petersburg

If the overall warming and decrease of ice cover continues, the development of transit shipping between European and South-East Asian countries can be expected. But scientists have different views on the probability of future warming.

In any case, exploration of Arctic shelf deposits of all types – first of all, hydrocarbons – will be extensively developed.

New icebreakers, mainly nuclear with 100–150 MWT capacity, will be built, providing year-round reliable and safe navigation, and offshore production platform operation.

Accordingly, the living standards of northern indigenous people will rise.

Martin Pratt, Director of Research, International Boundaries Research Unit, Durham University

My expectation is that in 2040 we shall still be looking at an incomplete jurisdictional map, and there will still be a debate about whether there is a need for an Arctic treaty, but that disagreements over who has rights over which areas will be addressed through diplomacy rather than through military confrontation – that is, no great change from today.

Claes Lykke Ragner (former Head of INSROP Secretariat), Fridtjof Nansen Institute

I am not an expert on ice, but believe it has been demonstrated beyond doubt that the extent and character of the arctic ice cover has changed greatly over the past decades and that it will probably continue to do so. Contrary to what I perceive to be popular belief and the main tone of media coverage, I do not, however, expect these changes to lead to a drastic increase in arctic shipping by 2040, nor do I foresee an Arctic beset by tension and insufficient regulation.

Regarding shipping, one tends to forget that although transit sailing routes may become ice-free for a few months every summer rather than a few weeks, the Arctic Ocean will still be covered in ice, and under extreme weather conditions, for most of the year. I believe this simple fact, along with the great and unpredictable annual variations in ice conditions, will be the main reason (among several others) why the Arctic Ocean will not have developed into an artery of intercontinental shipping by 2040, although a gradual but moderate increase of summer activities can be expected.

As for serious conflicts over jurisdiction and natural resources, I see little potential, since most activities and jurisdictional matters are already basically regulated under the *UN Convention on the Law of the Sea*. Four of the five arctic coastal states are parties to the convention, and the USA will in all likelihood become so long before 2040. The most important of the Arctic Ocean delimitation issues – the ongoing international process of determining the outer limits of the coastal states' continental shelves – will probably have been settled by then. In any case, there are no indications that the central areas of the Arctic Ocean have any great resource potential, and this limits the conflict potential of this process. Other jurisdictional disagreements, such as the delimitation between Norway and Russia in the Barents Sea, the status of the straits in the North-West and North-East passages and so on, might not be resolved by 2040, but it is difficult to imagine that such disputes will develop into something more serious. It should also be taken into consideration that the arctic states already have a long and good track record of bilateral and multilateral co-operation, be it on fisheries regulation, environmental protection or social, cultural and indigenous affairs, and should be able to extend that co-operation to new problem areas as they emerge.

In conclusion, in reply to the initial question, I foresee few drastic changes in the fields my institute has been following most closely. Even if there were to be major changes to the natural environment, I still see the Arctic of 2040 as an essentially pristine and inhospitable place, with a little bit more reasonably well-regulated shipping, petroleum activities and fishing along its fringes, and with neighbour relations that continue to be essentially good.

Mark Serreze, Director, National Snow and Ice Data Center, University of Colorado

It is very likely that we will have essentially no sea ice in summer. We will also likely be seeing the effects of these ice conditions on weather patterns, both within and beyond the Arctic.

Niels Stolberg, President and CEO, Beluga Shipping, Bremen

By 2040, the Arctic will have become an area of quite regular sea traffic, at least during summer. The still-growing trading centres in Europe and Asia will have played their part in establishing an efficient seaway through the Arctic Ocean, and those shipping companies which dispose of the necessary ice-hardened multipurpose heavy lift project carriers will be using the Northern Sea Route as a direct all-water connection between the continents quite frequently – as long as the window of opportunity allows. Super heavy lift modules will be shipped along the North Russian shore and used for the construction or remanufacturing of large power plants or refineries.

From an environmental perspective, we can only hope that by 2040 there will be arctic ice remaining, which can temporarily melt over the summer and thereafter freeze again. We hope that by using the Northern Sea Route, which also results in lower bunker consumption and emissions compared to the much longer transit through the Suez Canal, the global carbondioxide balance will look a fair lot better.

Peter Wadhams, Professor of Ocean Physics, University of Cambridge

By 2040, I think the Arctic will be completely ice-free in summer, so all three trade routes will be in operation, oil and mineral exploration will be widespread, polar bears will be in a bad way and the whole northern hemisphere will have reacted in unpredictable ways to this big change.

A Chronology

981–6	Viking Erik the Red explores and settles in Greenland
1001	Leif Erikson explores 'Vinland', in North America
1492	Columbus finds the West Indies
1497	Cabot reaches North America
1553	English merchant adventurers attempt N-E Passage
1576	Frobisher attempts N-W Passage
1596	Barents discovers Spitsbergen
1607	Hudson reports Spitsbergen coast teeming with whales
1610	Hudson enters his bay (and is murdered there in 1611)
1616	Baffin crosses his bay to enter the N-W Passage
1670	Hudson Bay Company monopoly granted by King Charles II
1697	Cossacks reach the Pacific
1703	Peter the Great's forces reach Baltic near modern St. Petersburg
1733–41	Great Northern Expedition maps coastline of Siberia
1778	Cook confirms separation of Asia and North America
1817	Mary Shelley creates the story of 'Frankenstein'
1818	Ross enters the N-W Passage at Lancaster Sound, but turns back
1819	Parry discovers channel deep into N-W Passage
1831	Ross (James Clark) locates the North Magnetic Pole
1837	First steam-powered icebreaker operates in Philadelphia, USA
1845	Franklin departs on fatal voyage

1847	Death of Franklin
1850–5	Collinson and McClure discover N-W passage
1858–9	McClintock's expedition finds relics of Franklin's men
1864	First Russian icebreaker, on River Neva in St. Petersburg
1869	Suez Canal opens
1876	First Canadian icebreaker ferries across Northumberland Strait
1878–9	Nordenskiold makes first transit of N-E Passage from W to E
1891–1904	Trans-Siberian Railway built
1893–6	Nansen's *Fram* drifts across polar ice cap
1898	*Yermak*, first sea-going icebreaker, launched at Newcastle-upon-Tyne
1903–6	Amundsen navigates N-W Passage
1905	Naval battle of Tsushima between Russia and Japan
1913	Vilkitsky explores N-E Passage – discovers Severnaya Zemlya
1914	Panama Canal opens
1915	Vilkitsky makes first E-W transit of N-E Passage
1916	Murmansk established
1917	Russian revolution
1920	Spitsbergen (Svalbard) Treaty signed – without USSR
1920	Canadian arctic oil strike on MacKenzie River
1926	Amundsen, Ellesworth and Nobile cross North Pole in airship
1929	Igarka Siberian timber port established
1930	Canadian Hudson Bay port of Churchill opens
1932	Soviet icebreaker transits N-E Passage transit in one season
1932	Stalin establishes Glavsevmorput
1935	USSR signs Spitsbergen Treaty
1936	Montreux Convention controls naval access to Black Sea
1937	Papanin pioneers North Pole ice-drifting station

A Chronology

1937–40	*Sedov* drifts across polar ice cap
1939	Murmansk Shipping Company established
1940	German raider *Komet* transits N-E Passage
1941	Murmansk besieged by German forces
1948	IMCO, now International Maritime Organisation, established
1949	NATO forms
1951–3	US Thule air base constructed in northern Greenland
1954–7	DEW Line early warning radar chain built across Arctic
1955	*Nautilus*, first nuclear-powered submarine, commissioned
1955	Admiral Gorshkov takes command of Soviet Navy
1956	Nasser nationalises Suez Canal
1957	*Lenin*, nuclear-powered icebreaker, launched
1958	USS *Nautilus* makes first underwater transit of North Pole
1959	USS *Skate* surfaces at North Pole
1960	USS *Seadragon* makes underwater transit of N-W Passage
1961	'Czar Bomba' 50-megaton nuclear test on Novaya Zemlya
1962	Cuban missile crisis
1963	Soviet submarine *Leninsky Komsomol* surfaces at North Pole
1964	Glavsevmorput dismantled
1964–90	Underground nuclear testing on Novaya Zemlya
1967	Soviet government invites foreign ships to use NSR
1967	Nuclear icebreaker *Lenin* 'disappears'
1968	Prudhoe Bay oil field discovered in Alaska
1969	Canadian icebreaker *Louis S. St-Laurent* commissioned
1969	Icebreaking tanker *Manhattan* transits N-W Passage
1970	*Lenin* returns to service
1970	Canada introduces Arctic Waters Pollution Prevention Act
1975	Suez Canal reopens

1977	Oil starts pumping through Trans-Alaskan Pipeline
1977	Nuclear icebreaker *Arktika*, first surface vessel to reach North Pole
1978	Year-round navigation of NSR commences, between Murmansk and Dudinka
1978	USCG icebreaker *Polar Sea* commissioned
1982	*UN Convention on the Law of the Sea* opens for signature
1984	Snohvit gas field discovered in Barents Sea
1985	USCG icebreaker *Polar Sea* transits N-W Passage
1985	Canada's arctic archipelago declared to be 'internal waters'
1987	Gorbachev renews invitation to foreign ships to use NSR
1988	Shtokman gas field discovered in Barents Sea
1989	*Exxon Valdez* oil spill in Alaska
1989	Berlin Wall falls
1991	Northern Sea Route formally open to foreign shipping
1991	Soviet Union disintegrates
1994	USA signs (but does not ratify) *UNCLOS*
2000	Putin becomes President of Russian Federation
2000	Russian submarine *Kursk* sinks in Barents Sea
2001	Russia submits first claim to *UNCLOS* continental shelf commission
2006	Norway submits claim to *UNCLOS* commission
2006	*Norilsky Nickel* double-acting ship commissioned
2007	HMS *Tireless* suffers explosion under arctic ice
2007	Extent of Arctic ice cap shrinks to record low
2007	Russian aircraft carrier *Admiral Kuznetsov* returns to high seas
2007	Russian freighter *Kapitan Sviridov* opens the 'Arctic Bridge'
2007	Norway's Snohvit gas field begins production

A Chronology

2008	Russian nuclear icebreakers transferred from MSCO to Rosatom
2008	N-E and N-W passages simultaneously open
2008	Medvedev becomes president of Russian Federation
2008	President Medvedev declares Arctic 'region of strategic importance'
2008	Russian Navy resumes arctic patrols
2008	Greenland votes for self-government
2008	Canada announces plans for polar icebreaker
2009	Russia shuts off Ukrainian gas supply
2009	New Russian security strategy suggests Arctic area of potential conflict
2009	*Lenin*, nuclear icebreaker, becomes museum
2013	Earliest forecast of ice-free summer arctic
2013	Deadline for Canadian submission to *UNCLOS* commission
2014	Deadline for Danish submission to *UNCLOS* commission
2014	Enlarged Panama Canal to open
2015	New generation of Russian icebreakers to enter service
2015	Nanisivik Canadian arctic naval base to open
2030–40	Arctic Ocean forecast to be free of summer ice

INDEX

A
Aida, Verdi's opera, 87
Alaska
　Exxon Valdez oil spill, 131, 153
　oil pipeline, 120, 130, 136
　Prudhoe Bay oilfield, 129, 136
Allen, Admiral Thad, 119
AMSA (Arctic Marine Shipping Assessment 2009), 138, 153–154, 189
Amundsen, Roald, 36, 38, 84
Anian, Strait of, 125
Antonov, Aleksander, 172–173
ARCOP (Arctic Operational Platform Project, 2003–6), 152, 163
Arctic
　animals, 5, 17, 40
　climate, 6, 7, 12, 175
　convoys, 48, 139
　cruises, 155, 190–191
　fish, 19
　flowers, 4
　gas, 20–22, 193
　ice cap, 3, 176–178
　minerals, 24, 115, 120, 136, 138
　oil, 20–22, 117
　taiga, 6, 16, 175
　tundra, 4, 6, 176
　whales, 5, 17, 138, 192

'Arctic Bridge', 139
Arctic Circle, 2, 126
Arctic Climate Impact Assessment, 12, 175
Arctic Council, 2, 153
Arctic Ocean
　currents, 9, 10
　description, 2, 10, 11
　fisheries, 19, 176
　ice extent, 6, 13, 176–178, 194
　navigable routes, 6, 13
　resources, 15–25
　weather, 23, 194–195
'Arctic Pilot Project', 137
Arkhangelsk, 16, 97, 101, 161
Armstrong, Terence, 49, 96
　See also Acknowledgements

B
Back, George, 75
Back, river, 75, 83–84
Baffin, William, 69
Bakayev, Viktor, 99, 100
Barber, Professor David, 197–198
Barents Sea, 2, 6, 66, 101–102, 147
Barents, Willem, 65–66
Barrow, John, Admiralty Secretary, 72
Barrow Strait, 79, 81
Baruch, Bernard, 47

Basque whalers, 18
Bathurst Inlet, 136
Bear Island, 65
Beaufort Gyre, 10
Beaufort Sea, 130, 136–137
Beaver, 17
Bellona, 106
Beluga Foresight, ship, 173
Beluga Fraternity, ship, 172
Beluga Shipping Group, 171–172
   See also Acknowledgements
Beluga white whale, 5
Bennett, Simon, 198–199
   See also Acknowledgements
Bering Strait, 2, 11, 72, 86
Bering, Vitus, 70–71
Bernstein, William, 88
Blackwood, John, 83
'Blubbertown' (Smeerenburg), 18, 40
Boothia Peninsula, 75, 79
Booth, Sir Felix, 64, 74
Bowhead whale, 5
BP, *see* Acknowledgements
Brent goose, 5
Brigham, Professor Lawson, 153, 199–200
   See also Acknowledgements
Bylot, Robert, 69

C

Canada
   Archipelago, 125
   Arctic Waters Pollution Prevention Act 1970, 132
   Bathurst Inlet, 136
   claims under UNCLOS, 31, 36–38, 137

icebreakers, 121–122, 134
military developments, 56, 135
Nanisivik, 56, 135
Nunavut, 135–136
oil and gas, 137–138
Resolute, 135
St. Lawrence Seaway, 121
shipping controls, 31, 133–134
sovereignty, 121, 130–133
Canadian Ice Service, 181
Canal du Midi, 88
Caribou (reindeer), 136, 176
Catherine the Great, Czarina, 52
Chernobyl, nuclear accident, 103, 105
Chilingarov, Artur, 26, 37–38
Chukchis, 70–71
Churchill, port of, 139
Circum-Arctic Resource Appraisal (USGS 2008), 20–21
Clark, Joe, Canadian Secretary for External Affairs, 125, 133
CLCS (Commission on the Limits of the Continental Shelf), 38
Climate change
   arctic 'amplifier', 12
   effect on Gulf Stream, 7, 183–185
   ice forecasts, 13, 18, 187
   methane release, 7, 183
   permafrost, 7, 176, 183
   temperatures, 12, 175
   'tipping points', 12, 183, 196
   weather effects, 194–195
CND (Campaign for Nuclear Disarmament), 99, 104

CNIIMF (Central Marine Research and Design Institute), 114, 156, 160
Coal, 15, 40
Cod, arctic, 19
'Cod wars', 18
Cold War, 15, 29, 44, 47, 99, 101, 188
Collinson, Richard, 81–82
Container ships, 152, 169, 196
Continental shelf, see UNCLOS (United Nations Convention on the Law of the Sea)
Cook, James, 72, 79
Coppermine, river, 73, 75
Cossacks, 70
Cruise ships, arctic, 155, 190–191
Cuban convoys, 19, 48, 99

D
Davis, John, 67–68
Dawkins, Richard, 28
Delgado, James P., see Acknowledgements
Denmark
  arctic claims, 35–36, 38–39
  Greenland's status, 22, 135, 138
  Hans Island, 36
  Thule air base, 48
DEW (Distant Early Warning) Line, 47
Dickens, Charles, 81
Dikson, 97
Dogger Bank incident, 50
Dome Petroleum, 137
Double-acting ships, 116
Drifting ice stations, 95, 98

Dudinka, 115, 147
Durov, Valeriy, 173

E
East Siberian Sea, 147
EEZ (Exclusive Economic Zone), see UNCLOS (United Nations Convention on the Law of the Sea)
Elizabeth I, Queen, 5, 63–64
Ellesmere Island, 39
Eskimos (Inuit), 75, 80–81, 85, 126–127
EU (European Union) energy policy, 24–25
Exxon Valdez, tanker, 131, 153

F
Felixstowe, Port of, see Acknowledgements
Finland
  Aker Arctic Technology, 117
  icebreakers, design and construction, 108, 113, 117
  territory lost, 101
Fishing
  arctic cod, 19
  disputes, 18, 19
Fox, arctic, 5, 6, 17
Fram (Nansen's), 9, 96
Fram Strait, 10
Franklin, Jane, 76, 78, 80, 82
Franklin, John
  1845 expedition, 76
  London statue, 83
  overland expedition, 74
  search for, 80–83

213

Franz Josef Land, 9, 35
Frobisher, Martin, 63
Frolov, Alexander, 195
Fur trade, 17

G
Gas
  arctic distribution, 21–22
  European dependence on, 24
  'gas wars', 25
  LNG tankers, 23, 137, 161, 194
  Shtokman field, 22–23, 193
  Snohvit field, 22
Gazprom, 23, 161
Gjoa (Amundsen's), 84
Glavsevmorput
  establishment, 93
  operations, 93, 95, 98, 145
  Stalin's purge, 95
Global warming, see Climate change
Godfrey, Peter, see
  Acknowledgements
Gorbachev, Mikhail, 101, 104, 141
Gorshkov, Admiral Sergei, 53–54, 99
GPS (satellite navigation), 195
Great Northern Expedition, 71
Greenland, 22, 64, 135, 138
Greve, Tim, see Acknowledgements
Gulag, Soviet, 95, 148

H
Hamburg, 170
Hans Island, 36
Harper, Canadian PM Stephen, 122
Herschel Island, 131
Herschel, William, 1
Hillier, Grant, see Acknowledgements
Hong Kong, 167

Hudson Bay, 68, 139
Hudson Bay Company, 69, 70
Hudson, Henry, 17, 67–68
Humble oil company, 130

I
Ice
  dangers, 8, 96
  extent, 6, 13, 176–178
  forecasts, 13, 194
  ice-strengthening, 169
  measuring, 3, 62, 176, 182
  melting process, 6, 12
  thickness, 3
  transport, 7–8
  types of, 3, 8, 9
Icebreakers
  costs, 111, 118, 148
  design, 110–111
  double-acting, 116
  future requirements, 114
  history, 107–108
  NSR tariffs, 170
  nuclear-powered, 109, 110–113
  operation, 111
Icebreakers, individual
  Arktika, 111, 118
  Chelyushkin, 94
  Fyodor Litke, 94, 110
  Healy, 119
  John G. Diefenbaker, 123
  John A. McDonald, 129
  Krasin, 119
  Lenin, 109, 110
  50 Let Pobedy, 112, 118, 172
  Louis S. St-Laurent, 122
  Norilsky Nickel, 117
  Polar Sea, 119, 120–121, 131, 133

INDEX

Polar Star, 119, 120
Rossia, 11, 172
Sedov, 8–9, 96
Sibir, 112, 143
Sibiryakov, 93, 108
Taimyr, 113
Vaygach, 113
Yamal, 106
Yermak, 95, 107
Iceland, 18, 33, 48, 196
Igarka, 16, 95, 144, 149
IMO (International Maritime Organisation), 27, 152–153
INSROP (International Northern Sea Route Programme), 149–150, 163, 171
International|Seabed Authority, 35
Inuit (Eskimos), 75, 80–81, 85, 126–127, 135
Ivan the Terrible, Czar, 16

J
Jackson, Frederick, 10
Japan, 50, 101, 143
Johansen, Frederik, 9
Juurmaa, Kimmo, 200
   See also Acknowledgements

K
Kamchatka, 70
Kara Gates, 147
Kara Sea, 16, 65, 102–103, 105, 145
Khodorkovsky, Mikhail, 24
King William Island, 79, 80, 82, 84
Kiriyenko, Sergei, 114
Kirkenes, 150

Kola Bay, 106, 161
Kola Peninsula, 48, 101
Krushchev, 53, 103
Kursk, submarine, 59

L
Lambert, Andrew, 81
   See also Acknowledgements
Lancaster Sound, 69, 73, 135
Laptev Sea, 85, 96, 146–147
Latvia, 148, 151
Lavrov, Sergei, Russian Foreign Minister, 38
Law of the sea, see UNCLOS (United Nations Convention on the Law of the Sea)
Lena, river, 144
Lenin, nuclear icebreaker
   design, 109
   'disappearance', 110
   museum, 118
Lewis, Peter, see Acknowledgements
Lomonosov, Mikhail, 141
Lomonosov Ridge, 11, 35, 39, 98
Longyearbyen, 42–43
Lukoil, 117, 159

M
MacKay, Peter, Canadian Foreign Minister, 37
Mackenzie, Alexander, 137
Mackenzie, river, 73, 137
Magadan, 148
Magnetic North Pole, 2, 69, 74–75, 85
Manhattan, tanker, 128–130
McClintock Channel, 139
McClintock, Leopold, 82–83

215

McClure, Robert, 81
McClure Strait, 73, 81, 127, 129, 137
McGoogan, Ken,
    see Acknowledgements
Medvedev, Dimitry, Russian
    President, 23, 55, 158
Melville Island, 73, 137
Mercator, 68
Mercator's projection, 165
Mirsky, Jeannette,
    see Acknowledgements
Mir, submersible, 37
Molotov-Ribbentrop pact, 97
Montreux Convention 1936, 54
Mumbai, 90
Murmansk, 6, 52, 97, 101, 115,
    139, 141
Murmansk Shipping Company, 102,
    113–114, 140, 156, 160

N
Nanisivik, 56, 135
Nansen, Fridtjof, 9–10
Nantucket, 17
Narwhal whale, 5
NATO, 41
Nautilus, US submarine, 60
N-E Passage
  Beluga voyages, 172–173
  distance savings, 167
  exploration, 64, 71
  future, 149–150, 158
  ice conditions, 146–147,
    178–179
  ice-free years, 178–179
  navigational routes, 147
  shipping traffic, 148–149,
    154–155
  transits, 151, 171–174
  weather, 146
Nicholas II, Czar, 50
Nordenskiold, Baron, 85–86, 93
Nord Stream pipeline, 23
Norilsk mining combine, 115
Norilsky Nickel, double-acting
    ship, 117
Northern Sea Route (Sevmorput)
  administration, 93, 95, 147, 159
  alternative routes, 147
  definition, 144–145
  development, 100, 158, 160,
    162, 164
  distance savings, 167
  Glavsevmorput, 93, 95, 98
  icebreaking support, 156, 160,
    162–163, 174
  navigational problems, 145–147
  opening to foreign ships, 98–99,
    100, 143, 148, 158
  rivers, 113, 144–145
  shipping traffic, 148–149, 162
  statistics, 143
  studies, 149–150, 152, 163, 171
  weather, 146
North Pole
  climate, 3
  transits, 60
  voyages to, 11, 37, 111, 143
Norway
  arctic claims, 35, 45
  fisheries, 19, 35–36
  safety concerns, 104, 191
  Snohvit gas field, 22
  See also Spitsbergen (Svalbard)
Novaya Zemlya, 65, 99,
    103–105, 147

Novosibirskiye Islands (Ostrova), 11, 39, 147
Novy Port, 144, 172
NSIDC (National Snow and Ice Data Center), Colorado, 177–178, 180, 203
Nuclear
   accidents, 103, 109, 110
   bomb testing, 103–104
   pollution, 103–104
   propulsion, 58, 60, 109, 111
Nunavut, 135–136
N-W Passage
   discovery, 80–82, 84
   distance savings, 167
   exploration, 64, 67, 69, 72
   geography, 125
   ice conditions, 138–139, 177–178, 180–181, 187–188
   ice-free years, 6, 138, 177, 179
   legal status, 131–134, 188
   navigational routes, 127, 138
   shipping traffic, 135, 138, 180

O
Obama administration – UNCLOS, 30
Ob, river, 105, 144
Oil
   arctic reserves, 20–22
   Prudhoe Bay field, 129, 136–137
   spills, 152–153, 190–191
Orwell, George, 47

P
Palin, Sarah, 120, 192
Panama Canal
   construction, 91
   distance savings, 92
   future development, 92
   opening, 91
Papanin, Ivan, 95–96
Parry Channel, 73, 127, 179, 181
Parry, Edward, 73, 81
Patrushev, Nicolai, 55
Patterson, Bill, 184
Peresypkin, Dr V. I., 200–201
Permafrost, 7, 176, 183
Peter the Great, Czar, 52, 70
Pevek, 147
Piracy, 168
Plutonium, 104
Polar bears, 5, 7, 42
Polar ice cap, 3
   See also Ice; Arctic Ocean
Polar Sea, icebreaker, 119, 120–121, 131, 133
Pole Star, 1
Poppy, arctic, 4, 15
Pratt, Martin, 201
   See also Acknowledgements
Prince of Wales Strait, 81, 127, 129
Prirazlomnoye oil terminal, 117, 161
Prudhoe Bay oilfield, 129, 136–137
Putin, Vladimir, Russian PM, 23–24, 37, 55, 114, 158

R
Rae, Dr John, 80–81
Ragner, Claes, 201–202
Reindeer (caribou), 4, 136, 176
Resolute, 135
Richardson, John, 84

Robinson, Kate, *see* Acknowledgements
Rosatom, 114, 160, 163
Ross, James Clark, 75, 79
Ross, John, 73–74, 79
Rotterdam, 167
Rozhestvensky, Admiral, 50
Russia Company, 16–17, 68
Russia
  arctic claims, 35, 38–39, 44
  arctic resources, 16–17, 19, 21–23
  arctic strategy, 55–56, 117, 158–159
  economy, 22, 159, 189, 193
  fisheries, 19
  indigenous peoples, 70, 157
  naval policy, 52–58
  nuclear industry, 103
  oil and gas, 22–24, 161
  revolution, 53, 93
  shipping policy, 100
  See also Northern Sea Route (Sevmorput)

S
Sabine, Edward, 73
St. Petersburg, 52, 93, 101
Sakharov, Andrei, 104
San Francisco, 18
Satellite surveys, 3, 6
Scoresby, William, 72
Scott Polar Research Institute, Cambridge, 96
Scott, Robert, 83, 85
Sedov, icebreaker, 8–9, 96
Serreze, Mark, 178, 203
  See also Acknowledgements
Severnaya Zemlya, 4, 146–147, 181

Severomorsk, 58, 102
Shanghai, 167
Shelley, Mary, 1, 2, 10, 196
Ships, individual surface
  Admiral Kuznetsov, 57
  Admiral Scheer, 49, 97
  Ancon, 91
  Aurora, 53
  Charles Hanson, 85
  Chelyuskin, 94
  Erebus HMS, 78, 82
  Exxon Valdez, 131, 153
  Fox, 82
  Fram, 9
  Fury, HMS, 74
  Gjoa, 84
  Hansa, 99
  Indiga, 106
  Isabella, 75
  Jeannette, 9
  Kapitan Sviridov, 139–140
  Komet, 49
  Manhattan, 128–130
  Marienburg, 99
  Novovoronezh, 101
  Otto Hahn, 108
  Peter the Great, 57
  Savannah, 108
  Sovietsky Soyus, 99
  Terror HMS, 78, 82
  Vasily Dinkov, 117
  Vega, 85–86
Shmidt, Otto, 95
Shtrek, Alexey,
  see Acknowledgements
Siberia, 6, 70
Sibiryakov, icebreaker, 93, 97
Simpson, Thomas, 76, 84

Smeerenburg ('Blubbertown'), 18
SOLAS (Safety of Life at Sea)
  convention, UN, 27
Sonar, 61
Sovcomflot, 117, 162
Spitsbergen (Svalbard)
  airfield, 42
  Barentsburg mine, 15
  discovery, 65
  Longyearbyen, 42–43
  Russian presence, 40, 42
  scientific research, 44
  Starostin, Ivan, 40
  strategic importance, 44
  Svalbard poppy, 15
  Svea mine, 15
  Treaty, 34, 39, 40, 44–45
  whaling, 17
Stalin, 53–54, 96, 157
Starvation Cove, 84
StatoilHydro oil company, 23
Stolberg, Niels, 173–174, 203
Submarines
  accidents, 59
  arctic transits, 60
  nuclear propulsion, 58, 60
  operations under ice, 58
  strategic importance, 58
Submarines, individual vessels
  Alexandria, 61
  Kursk, 59
  Leninsky Komsomol, 61
  Mir, 37
  Nautilus, 60
  Skate, 60
  Tireless, 61–62
Suez Canal
  construction, 89

de Lesseps, Ferdinand, 89
Disraeli, Benjamin, 89
distance savings, 87, 90
Eden, Anthony, 89
history, 87–88
Nasser, Gamal Abdel, 89
tolls, 90, 167, 193
Verdi's opera Aida, 87
wartime closures, 89–90, 99
Svalbard poppy, 15
Svalbard, see Spitsbergen
  (Svalbard)
Sweden, 40, 52, 191

T
Taiga, 6, 16, 175
Tankers
  azipod, 117
  ice-strengthened, 117–118
  LNG (liquid natural gas), 23, 117,
    137, 161, 194
  Manhattan, 128–130
  shuttle, 117, 161
  Vasily Dinkov, 117
Taylor, David, see
  Acknowledgements
Temperatures, global
  average, xxxi
Thule air base, 48
Timber trade, 16
Tireless, HM submarine, 61–62
Total oil company, 23
Transpolar Drift, 10
Trans-Polar Route, 181, 194
Trans-Siberian Railway, 16, 144
Tromso, 41, 167
Tsushima, battle of, 51
Tundra, 4, 6, 176

U

UNCLOS (United Nations Convention on the Law of the Sea)
  'arctic exception', 132
  Article, 132, 234
  baselines, 31, 133
  continental shelf, 32
  deadlines, 38, 193
  EEZ (Exclusive Eonomic Zone), 33
  'freedom of the seas', 31
  internal waters, 31, 133, 147
  right of innocent passage, 32, 147
  right of transit passage, 32
  territorial waters, 31, 148
  US attitudes, 30
USA (United States)
  Alaskan oil, 22, 120, 129–131, 136–137
  icebreakers, 107, 118–121
  legal claims, 36, 38, 137
  National Research Council, 120
  naval policy, 30, 119, 132
  US Coast Guard, 119–120
US Geological Survey, 20

V

Vancouver, 167
Varandey oil terminal, 117
Variation, magnetic, 2
Vasilyev, Vladimir, see Acknowledgements
Vikings, 64
Vilkitsky, Commander, 51, 108
Vilkitsky Strait, 146, 171–173, 178
Vladivostok, 50, 52

W

Wadhams, Professor Peter, 62, 184, 203
See also Foreword; Acknowledgements
Walter, Dr Katey, 183
Whalebone (baleen), 17–18
Whales
  beluga, 5
  bowhead, 5
  human uses, 17
  narwhal, 5
  navigation, 138
  right, 5, 17
  whaling boom, 17
Winter Harbour, 81
Wrangel Island, 35
Wright, Peter, see Acknowledgements
Wunderland, operation, 97

Y

Yenesei, river, 16, 105, 144
Yermak, icebreaker, 95, 107
Yermak, Timofeyevich, 70
Yokohama, 98, 150, 167, 170
Yukon, 36

Z

Zeno map, 67